Techniques and Sample Outputs that Drive Business Excellence

The **LITTLE BIG BOOK** Series

Other books in THE **LITTLE BIG BOOK** SERIES

Performance Acceleration Management, H. James Harrington (2013)

Closing the Communication Gap, H. James Harrington and Robert Lewis (2013)

Maximizing Value Propositions to Increase Project Success Rates, H. James Harrington and Brett Trusko (2014)

Making the Case for Change: Using Effective Business Cases to Minimize Project and Innovation Failures, H. James Harrington and Frank Voehl (2014)

Techniques and Sample Outputs that Drive Business Excellence, H. James Harrington and Charles Mignosa (2015)

Recreating Information Technology: As a Catalyst to Deliver Total Organizational Change, H. James Harrington, Richard Harrington Jr., and Ron Skeddle

Creativity, Innovation and Entrepreneurship: The Only Way to Renew Your Organization, H. James Harrington, Richard Harrington Jr., and Ron Skeddle

Rules of the Road for Entrepreneurs: A Map for a Successful Journey, H. James Harrington and Shane P. Rogers

Organizational Change Management, H. James Harrington

Project Management, Review and Assessment, H. James Harrington and William S. Ruggles

Cementing Customer Relationships, H. James Harrington and Thomas H. Carmody

Organizational Portfolio Management for Projects and Programs, H. James Harrington, Richard Harrington Jr., and Ron Skeddle

Using Software Technology to Improve Organizational Performance, H. James Harrington and William S. Ruggles

Techniques and Sample Outputs that Drive Business Excellence

H. James Harrington
Chuck Mignosa

CRC Press is an imprint of the
Taylor & Francis Group, an **informa** business
A PRODUCTIVITY PRESS BOOK

CRC Press
Taylor & Francis Group
6000 Broken Sound Parkway NW, Suite 300
Boca Raton, FL 33487-2742

Printed on acid-free paper
Version Date: 20141205

International Standard Book Number-13: 978-1-4665-7726-8 (Paperback)

Visit the Taylor & Francis Web site at
http://www.taylorandfrancis.com

and the CRC Press Web site at
http://www.crcpress.com

I dedicate this book to George and Sandy Weiner. I'm so thankful to have them as good friends and grandparents to my granddaughter Grace. I couldn't have found two better people.

H. James Harrington

This book is dedicated to Eileen, the person who gave me the space to be crazy and supported my insane explorations into the eclectic universe of the mind.

Chuck Mignosa

Contents

Preface

Many people and organizations are looking for that magic tool or methodology that will suddenly transform them into an outstanding organization. Consultants from every part of the world preach that they have the answer for you. Professional organizations of all kinds are blessed with the unique ability to define that one approach or method that can erase all your problems with a minimal amount of effort. Your controller has that magic answer on how to reduce costs. The manufacturing engineering department insists that the problem you are facing is low productivity and they have the answer to that. The quality assurance department is insistent that their program, Lean Six Sigma, will completely eliminate all the waste. The human resources (HR) manager points out that the only advantage the organization has is its skilled workforce and as a result, he/she insists that investing heavily in it is the key to future prosperity. The development engineering department claims that increased innovation is the only way to stay ahead of the competition and grow profits.

Everyone claims to have that silver bullet that will kill the dragons and vampires, eliminating all of your problems with one quick wave of that magic wand. Well, as one of the people who have lived through many of these cure-alls, I firmly believe there is no one right answer for all organizations or for even a single organization. The successful organization carefully combines these improvement approaches together to make a unique improvement "soup" that is tasty to all of its stakeholders. And, as important as the tools and methodologies can be, the honesty, commitment, and constancy of purpose at all levels of the management team are even more critical than what you put into your soup. We have seen these same tools and methodologies applied to organizations where, in one case, the return on investment was 40 to 1 and, in another case, where the same activity/project was a complete failure. Most of these tools, techniques, and methods are very successful when they are applied to the right situations and backed up by a management team that wants them to work and that is dedicated to make it happen.

Because many of the more common tools and methodologies are used in many of the approaches that are presented in the Little Big Book series, we have decided to prepare one book that discussed these common tools thereby eliminating the need for the reader to go over the same materials in the individual books where the tools/methodologies are used. As a result, this book can be used as a reference primer for all of the Little Big Book series as well as many of the other books that focus on how to improve an organization's performance. It is our hope that this approach will allow the reader of each book to stay focused on the approaches being presented without being sidetracked into learning all of the different performance improvement tools/methodologies that are actually used when they are being implemented.

H. James Harrington

Acknowledgments

I'd like to acknowledge all of the diligent effort Candy Rogers put forth to bring this book together and to keep it organized over the past months. She reacted courageously to the many changes in direction and content that I made in bringing it all together. Her never-give-up attitude inspires us all.

H. James Harrington

I'd like to acknowledge H. James "Jim" Harrington who allows me the space to argue about the simple and complicated details of life and, from the dust, discover new ways of looking at the world. They prove over and over again that one doesn't have to be wrong for another to be right and that belief systems blind us to reality. Thank you, H. J.

Chuck Mignosa

About the Authors

H. James Harrington, PhD, is the CEO of Harrington Institute, Inc., in Los Gatos, California. In the book, *Tech Trending*, Dr. Harrington was referred to as "the quintessential tech trender." *The New York Times* referred to him as having a "… knack for synthesis and an open mind about packaging his knowledge and experience in new ways—characteristics that may matter more as prerequisites for new-economy success than technical wizardry. …" The author, Tom Peters, stated, "I fervently hope that Harrington's readers will not only benefit from the thoroughness of his effort, but will also 'smell' the fundamental nature of the challenge for change that he mounts." President Bill Clinton appointed Dr. Harrington to serve as an Ambassador of Good Will. It has been said about him, "He writes the books that other consultants use."

The Harrington Institute was featured on a half-hour TV program, *Heartbeat of America*, which focused on outstanding small businesses that make America strong. The host, William Shatner, stated, "You (Dr. Harrington) manage an entrepreneurial company that moves America forward. You are obviously successful."

Dr. Harrington is recognized as one of the world leaders in applying performance improvement methodologies to business processes. He has an excellent record of coming into an organization as its CEO or COO, resulting in a major improvement in its financial and quality performance.

Dr. Harrington is a very prolific author, publishing hundreds of technical reports and magazine articles. He has authored 40 books and 10 software packages.

Experience

Dr. Harrington now serves as the CEO for the Harrington Institute with branches in many countries around the world. He also serves as the chairman of the board for a number of businesses.

In February 2002, Dr. Harrington retired as the COO of Systemcorp ALG, the leading supplier of knowledge management and project management software solutions in Toronto, Canada. Prior to this, he served as a principal and one of the leaders in the Process Innovation Group at Ernst & Young. Dr. Harrington was with IBM for over 40 years as a senior engineer and project manager.

He is past chairman and past president of the prestigious International Academy for Quality and the American Society for Quality Control. He is also an active member of the Global Knowledge Economics Council.

While he was chairman of ASQ, he was one of the leaders in getting the Malcolm Baldrige National Quality Award through the U.S. Congress and approved by the president. He also served as the first treasurer of the Malcolm Baldrige Consortium that set up and developed the award.

Credentials

Dr. Harrington was elected to the honorary level of the International Academy for Quality, which is the highest level of recognition in the quality profession. He is a government-registered quality engineer, a certified quality and reliability engineer by the American Society for Quality Control, and a permanent certified professional manager by the Institute of Certified Professional Managers. He is also a certified Master Six Sigma Black Belt and earned the title of Six Sigma Grand Master. Dr. Harrington has an MBA and PhD in engineering management and a BS in electrical engineering.

He was appointed the honorary advisor to the China Quality Control Association, and was elected to the Singapore Productivity Hall of Fame in 1990. He has been named lifetime honorary president of the Asia Pacific Quality Organization and honorary director of the Association Chilean de Control de Calidad. In 2006, Dr. Harrington accepted the honorary chairman position of Quality Technology Park of Iran. In 2008, Dr. Harrington was awarded the Sheikh Khalifa Excellence Award (UAE) in recognition of his "superior performance as an original Quality and Excellence Guru who helped shape modern quality thinking." In 2009, he was selected as the Professional of the Year (2009). Also in 2009, he received the Hamdan Bin Mohammed e-University Medal. In 2010, the Asian Pacific Quality Organization (APQO) awarded Dr. Harrington the APQO President's Award for his "exemplary leadership." The Australian Organization of Quality NSW's Board recognized him as "the Global Leader in Performance Improvement Initiatives" in 2010. In 2011, he was honored to receive the Shanghai Magnolia Special Contributions Award from the Shanghai Association for Quality in recognition of his 25 years of contributing to the advancement of quality in China. In 2012, Dr. Harrington received the ASQ Ishikawa Medal, the Jack Grayson Award, A.C. Rosander Award, and the Armand V. Feigenbaum Lifetime Achievement Medal. In 2014, Dr. Harrington was appointed chair of TQM College Manchester (U.K.) Board.

Dr. Harrington has been elected a Fellow of the British Quality Control Organization and the American Society for Quality Control. In 2008, he was made an Honorary Fellow of the Iran Quality Association and of the Azerbaijan Quality Association. He also was elected an honorary member of the quality societies in Taiwan, Argentina, Brazil, Colombia, Chile, and Singapore.

Recognition

- *The Harrington/Ishikawa Medal*, presented yearly by the Asian Pacific Quality Organization, was named after H. James Harrington to recognize his many contributions to the region.
- *The Harrington/Neron Medal* was named after H. James Harrington in 1997 for his many contributions to the quality movement in Canada.
- *Harrington Best TQM Thesis Award* was established in 2004 and named after H. James Harrington by the European Universities Network and e-TQM College.
- *Harrington Chair in Performance Excellence* was established in 2005 at the Sudan University (Khartoum).
- *The Harrington Excellence Medal* was established in 2007 to recognize an individual who uses the quality tools in a superior manner.

Charles (Chuck) Mignosa has more than 30 years of diversified experience in high technology, aerospace, telecommunications, food processing, and biomedical device industries.

He was with IBM for 25 years and holds patents in solid lubricants. He was the project manager in charge of implementing quality systems and is a certified course developer who developed many customized courses. Some examples include Total Quality Management, Continuous Flow Manufacturing, Customer Driven Quality, Statistical Design and Analysis of Experiments, Team Building, Six Sigma, and Communication Skills. Mignosa also was involved with IBM's Six Sigma projects that were designed after Motorola's program.

Mignosa has delivered courses in management development, team building, organizational excellence, conflict resolution, Stephen Covey's *7 Habits of Highly Effective People*, and principle centered leadership, along with many other topics.

After leaving IBM, he worked as an independent consultant, and has consulted for and done training with such companies as IBM, SanDisk, Owens Corning, Heinz USA, Siemens Automotive, General Mills, Connors Peripherals, HP, ADAC Labs, and many more. He has been an adjunct instructor with the University of Notre Dame (South Bend, Indiana), Nova University (Fort Lauderdale, Florida), and the University of Nevada/Las Vegas.

Mignosa spent three years as the director of quality for Cholestech Corporation (San Diego) where he brought them into FDA compliance and registered to ISO 9001 and ISO 13485. He also spent two years as vice president of quality at P-COM, a telecommunications equipment development company in Campbell, California, where he reduced out-of-box failures from 30 percent to less than 1 percent.

Mignosa presently is president and CEO of Business Systems Architects, LLC (Santa Clara, California). His teams specialize in the design and implementation of business and quality management systems, organizational excellence, continuous improvement Lean/Six Sigma, and strategic planning. His partners include H. James Harrington of the

Harrington Institute and Jerry Mairani of the Institute of Performance Improvement.

In addition to a BS in chemistry from San Jose State University, Mignosa has completed graduate work in statistics at Stanford, a masters in systems engineering at Systems Research Institute in New York, and management training with IBM. He is a senior member of the American Society for Quality (ASQ), and is a certified Master Black Belt trainer.

Introduction

If you are going to repair or correct anything, you need a good set of tools and you need to understand how to use them. This is true whether you are making repairs around the home, fixing your car, solving problems at work, or trying to find out why your daughter isn't speaking to you. Throughout this series of books, entitled *Little Big Books*, we have addressed a number of the problems and opportunities that individuals face in their everyday working environments. Most of these problems and opportunities make use of a common set of tools, but use them in different ways or combinations in order to effectively address and take advantage of the immediate conditions. It's a lot like doing work around the house. It doesn't matter if you are putting together a new bookshelf or replacing a defective light switch. In both cases, you need a screwdriver, a knife, and a pair of pliers. Because it is outrageously expensive to hire a plumber to fix the leaky faucet, a painter to paint that back bedroom, or a carpenter to install a corner bookshelf, we all turn into do-it-yourselfers. We all have a toolbox full of these critical tools that are required for us to meet these challenges in maintaining our homes and cars. Our toolboxes are full of different sized screwdrivers, wrenches, hammers, drills, paintbrushes, etc. A new electric drill is always an excellent gift to give oneself on Father's Day. I don't know about you, but I just love buying new tools for my toolbox. Every time I walk through Home Depot I see a new and shiny temptation that I want to add to the collection I already have. And, if I can't justify buying it for myself because I have something that already does that job, I can often justify buying it for my son and daughter-in-law because they surely need it. I already have a number of new and shiny tools that have never been dusted off or tasted fresh sawdust.

Just as it is important to have a toolbox full of the right tools at home, the same thing is true for all of us at work. I'm not talking about us trying to fix a light switches or repair the broken chair at work. I'm talking about the tools necessary to address our day-to-day problems and take advantage of the opportunities we have continuously presenting themselves to us through the process we are using in our normal working conditions. Just as we need a set of standard tools to make repairs around the house, we also need a set of standard tools to use every day at work to take advantage of the opportunities we have to improve the quality of the products we are delivering to our customers and the processes that we use to manage our organizations. These tools are not screwdrivers, pliers, or wrenches; they are tools like teambuilding, Pareto analysis, value analysis, business case analysis, brainstorming, benchmarking, etc. It doesn't matter whether you are installing a new customer relations management software package or restructuring the organization, you need to know how to effectively use the organizational change management methodology in order to ensure it is efficiently and effectively implemented. It doesn't make any difference which improvement group you attached your star to—Crosby, Deming, Juran, Peters, Ishikawa, Taguchi, etc. It doesn't matter what improvement methodology you are in the process of implementing—Total Quality Management, Lean, Six Sigma, Reengineering, Balanced Scorecard, Toyota Manufacturing, National Quality Award, Activity Based Costing, etc. It doesn't matter what types of teams you are using—Quality Circles, Performance Improvement Teams, Six Sigma Teams, Natural Work Teams, etc. It doesn't matter if you are trying to improve by restructuring or through the use of technology, acquisitions, etc. There is a set of basic tools that all of us need and depend upon in order to effectively and efficiently implement them. This book is about the basic screwdrivers, wrenches, hammers, saws, and drills that make up the basic tools in the performance improvement toolbox that every individual and every organization should have at their fingertips and have mastered the use of in order to be successful in today's complex business environment. These are the tools that are used in the everyday resolution of problems and to take advantage of improvement opportunities. They also are the tools that need to be readily available because it is impractical to stop and go out and buy them when the opportunity to use them arises because the delay in getting them often means that the organization loses the opportunity to take advantage of that improvement opportunity.

When we started this series of books, we debated if we should include the tools that were essential for the methodologies within the book itself. We decided not to include the basic tools in each of the individual books in order to reduce the amount of material that the individual reader would be subjected to and eliminate duplication from book to book. The basic tools that are included in this book are

- Affinity Diagrams
- Brainstorming (creative brainstorming)

- Cause-and-Effect Diagrams
- Check Sheet
- Commitment Building
- Consensus Building
- Consequence Management
- Control Charts
- Cost–Time Charts
- Data Gathering by Document Review
- Data Gathering by Interview
- Data Gathering by Samples
- Data Gathering by Surveys
- Data Stratification
- Delphi Narrowing Technique
- Employee Involvement Facilitated Sessions
- Facilitated Sessions
- Flowchart
- Force Field Analysis
- Graphs
- Kano Model
- Motivation Management
- Negative Analysis
- Nominal Group Technique
- Organizational Change Management
- Organizational Process Consultation
- Organizational Process Improvement (OPI)
- Pareto Analysis
- Prioritization Matrices
- Process Capability Analysis (Cp)
- Project Charter
- Project Management
- Risk Management
- Role Mapping
- Root Cause Analysis
- Run Chart (Statistical Process Control)
- Scatter Diagrams
- Solutions Evaluation
- Storyboarding
- Value Analysis (VA)
- Value Engineering

- Value Proposition
- Voting
- Workflow Diagram

In preparing to write this book, we needed to address a number of questions. One of them was how deep do we go into the use of an individual tool. Some of the individual tools are complex enough to justify a book solely dedicated to them. Typical examples of this type of tool are benchmarking and organizational change management. Other tools are straightforward enough that the application could be completely described in one or two pages. To solve this dilemma, we decided to address all tools to the point that the reader would have a working knowledge of the tool, but the more complex tools would not include a detailed understanding of all the ways the tool could be used.

Another question that we faced was how were we going to handle the new tools that could be added to the toolbox on a regular ongoing basis. When we analyzed this problem, we soon realized that these new tools were a lot like the tools we use around the home. There is a continuous flow of new products coming out, but most of them are a product that is replacing a red painted handle with a new shiny chrome-plated handle, a torque wrench drill bit holder replaced with a rapid release one, or a new drill with a battery that lasts 20 percent longer. Just as most of the new home improvement tools are just refinements of the old approaches, the same is true of most of the tools used to improve organizational performance. Six Sigma did not generate any new concepts. All of the statistical applications were well-defined and had been used for many years. The DMAIC (define/measure/analyze/improve/control) methodology was just a minor remodel of the plan–do–check–act methodology that was developed by Walter Shewhart back in the 1930s. As a result, the basic tools' application concepts and basic functions have changed very little over the years. Of course, refinements have been added and these are being implemented. And, of course, these refinements should be reflected in the way we are using these tools. We can't ignore things like the Internet, the higher degree of education the general population is receiving today, and the international competition with which every organization must deal.

It doesn't matter whether you are buying one, two, all, or none of the books in a Little Big Book Series. The tools that we are presenting in this book are the essential and basic tools that everyone in any organization should be using. As a result, the material presented here is relevant to the way your organization needs to function in today's competitive environment. We feel confident that by understanding and using these tools, you and your organization will receive significant benefit and added value.

H. James Harrington
Chuck Mignosa

1

Overview of the Approach Used in This Book

This book is designed to provide the reader with a general understanding of how the frequently used performance improvement tools/methodologies are used and what a typical output would look like. The way these tools/methodologies are presented here was designed to supplement and provide more detail in how to implement and use the common tools that are frequently referred to in the Little Big Book series and in the other books that are focused on improving organizational performance. Some of the tools that are discussed are very complex and, to be adequately addressed, it would require that a book be dedicated to just that tool in order for the reader to become a skilled user of the tool. A typical example of this is the methodology entitled "Design of Experiments." In other cases, the information presented in this book provides adequate direction to allow an individual to effectively use the tool in most applications. A typical example of this is "Brainstorming." The tools we have selected provide a combination of the tools that are used in problem solving, continuous improvement, and innovation. This approach allows the book to address the common tools/methodologies that are used in the books that make up the Little Big Book series.

This book presents, in alphabetic order, a total of 44 tools/methodologies each of which is presented in the following manner:

- Name of the tool/methodology.
- Definition of the tool/methodology.
- An explanation of what it is used for.
- An explanation of how to prepare and use the tool/methodology.

- An example of the completed output.
- A list of some of the software packages available to help in the use and implementation of the tool/methodology.
- Additional references to provide the reader with other sources of information related to the specific tool or methodology being discussed.

The following is a list of tools/methodologies and their definitions that are presented in this book:

1. **Affinity Diagrams:** A technique for organizing a variety of subjective data (such as options) into categories based on the intuitive relationships among individual pieces of information. It is often used to find commonalties among concerns and ideas.
2. **Brainstorming (creative brainstorming):** A technique used by a group to quickly generate large lists of ideas, problems, or issues. The emphasis is on quantity of ideas, not quality.
3. **Cause-and-Effect Diagrams:** A visual presentation of possible causes of a specific problem or condition. The effect is listed on the right-hand side and the causes take the shape of fish bones. This is the reason it is sometimes called a "Fishbone Diagram." It is also called an "Ishikawa Diagram."
4. **Check Sheet:** A simple form on which data are recorded in a uniform manner. The forms are used to minimize the risk of errors and to facilitate the organized collection and analysis of data.
5. **Commitment Building:** Commitment is a promise to give or do something, to be loyal to someone or something. It is the act of pledging or engaging oneself in an obligation or a promise to be engaged, or becoming involved in a given activity to achieve a given result.
6. **Consensus Building:** A technique to obtain the commitment of all team members to move in a particular direction.
7. **Consequence Management:** A formal process of understanding the institutional structures that reflect deeply held values and beliefs in the organization and then utilizing those structures (e.g., compensation or training opportunities) to influence desired behavior.
8. **Control Charts:** A graphic representation that monitors changes that occur within a process by detecting variation that is inherent in the process and separating it from variation that is changing the process (special causes).

9. **Cost–Time Charts:** A date-and-cost line chart that tracks a process changing cost over time. (Also referred to as a date-and-price chart, and similar to a Gantt chart.)
10. **Data Gathering by Document Review:** A technique to quickly collect information that currently exists within an organization.
11. **Data Gathering by Interview:** The act of using the interviewing process to collect data.
12. **Data Gathering by Samples:** A sample is a representation from a population that allows the observer to predict the actual distribution of the population. There is rarely enough time or resources to measure the whole population; a representative sample of the population will yield a model for the entire population.
13. **Data Gathering by Surveys:** Surveys are used to measure characteristics of a population relative to such things as behavior, awareness of programs, attitudes or opinions, and needs. Repeated surveys can give valuable information about trends, such as in evaluating government activities.
14. **Data Stratification:** A technique used to help identify the underlying causes of variation within a population of data.
15. **Delphi Narrowing Technique:** A tool that eliminates the need for face-to-face interaction while it enables achieving group consensus through the use of a prioritization scheme.
16. **Employee Involvement: A** technique for unleashing human potential in organizations and involving people in the change process.
17. **Facilitated Sessions:** A meeting in which the leader (facilitator) guides the discussions through a series of steps designed to arrive at a consensus that is acceptable to all participants. A facilitated session helps the participants to define and support mutual goals and objectives.
18. **Flowchart:** A method of graphically describing an existing or proposed process by using simple symbols, lines, and words to pictorially display the sequence of activities. Flowcharts are used to understand, analyze, and communicate the activities that make up major processes throughout an organization. They are essential tools used in Process Redesign, Process Reengineering, Six Sigma, and ISO documentation.
19. **Force Field Analysis:** A method to help identify the positive and negative forces working on a process when trying to attain a new state. It is a visual aid for pinpointing and analyzing elements that

resist change (restraining forces) or push for change (driving forces). This technique helps drive improvement by developing plans to overcome the restrainers and make maximum use of the driving forces.

20. **Graphs:** A method for visually comparing two or more sets of data. Graphs are visual displays of quantitative or qualitative data. They visually summarize a set of numbers or statistics.
21. **Kano Model:** A theory of product development and customer satisfaction, which classifies customer preferences into five categories:
    - Must-be quality
    - One-dimensional quality
    - Attractive quality
    - Indifferent quality
    - Reverse quality
22. **Motivation Management:** It is developing an understanding of what provides an individual with a sense of self-worth, pride, and accomplishment when using this understanding to guide the way work is assigned and how individuals are recognized for their efforts. It results in an organization developing its values, beliefs, procedures, and culture in a way that drives the organization's employees to get more enjoyment out of the work they are doing and be more committed to the organization. It makes use of a set of logical neuro programs that suit the perception of a person's or an organization's needs for the purpose of efficiency and optimality to accomplish desired organizational goal.
23. **Negative Analysis:** An approach used to look at a process or situation to define what action could be taken to cause a negative impact upon the results. It generates a list of actions that, if implemented, would result in making the present situation worse. It then generates action plans to minimize the impact that these actions would have upon the process or situation.
24. **Nominal Group:** A special purpose technique, useful for situations where individual judgments must be tapped and combined to arrive at decisions. It is a process to develop and narrow alternatives by generating ideas.
25. **Organizational Change Management:** A methodology designed to help prepare the organization and the individuals within the organization to accept changes to the organizational structure, processes, and operating procedures. It is designed to break down resistance to change and to build up organization resiliency.

26. **Organizational Process Consultation:** It is the combination of skill in establishing a helpful relationship, in knowing what kind of processes to look for, and in intervening in such a way that processes are improved.
27. **Organizational Process Improvement (OPI):** A combination of two methodologies: process reengineering and process redesign. It also is often called Business Process Improvement (BPI). OPI is a systematic approach to bringing about step function improvement in processes within an organization. It focuses on increasing adaptability, efficiency, and effectiveness while reducing cost and cycle time.
28. **Pareto Analysis:** A specialized type of column graph that is created to simplify comparisons between items.
29. **Prioritization Matrices:** A narrowing technique that is used to rank large lists of alternatives.
30. **Process Capability Analysis (Cp):** A statistical comparison of a measurement pattern or distribution to specification limits to determine if a process can consistently deliver products within those limits. It is a measure of the relationship between common system variation and the specification limit.
31. **Project Charter:** A document that formally organizes the project, thereby authorizing the project leader to enlist organizational resources to accomplish its objectives. It also defines what the project is responsible for accomplishing.
32. **Project Management:** The application of knowledge, skills, tools, and techniques to project activities in order to meet or exceed stakeholders' needs and expectations from a project. (Source: *Project Management Body of Knowledge* (PMBOK)® *Guide*)
33. **Risk Management:** Every project carries with it a certain amount of risk. Risk management is the process of identifying and prioritizing those risks, then both implementing strategies to manage them and designing contingency plans to supplement those strategies should the risk occur.
34. **Role Mapping:** A method of graphically defining the relationship of people relative to their professional relationships, political, and organizational structures that are necessary to the success of the various components of a major change.
35. **Root Cause Analysis:** The process of identifying the various causes affecting a particular problem, process, or issue, and determining the real reasons that caused the condition.

36. **Run Chart (Statistical Process Control):** A graphic display of data used to assess the stability of a process over time, or over a sequence of events (such as the number of batches produced). The run chart is the simplest form of a control chart.
37. **Scatter Diagrams:** A graphic tool used to study the relationship between two variables. Scatter diagrams are used to test for possible cause-and-effect relationships. It does not prove that one variable causes the other, but it does show whether a relationship exists and reveals the character of that relationship.
38. **Solutions Evaluation:** A technique used to help review and narrow solutions on the basis of a thorough cost/benefit analysis.
39. **Storyboarding:** The act of physically structuring the output into a logical arrangement. The ideas, observations, or solutions may be grouped visually according to shared characteristics, dependencies upon one another, or similar means. These groupings show relationships between ideas and provide a starting point for action plans and implementation sequences.
40. **Value Analysis (VA):** The act of identifying the required functions for a product, establishing values for the required functions, and suggesting an approach to provide the required functions at the lowest overall cost without performance loss to optimize cost performance.
41. **Value Engineering (VE):** A methodology that seeks to improve the "value" of goods or products and services by evaluating ways that costs or function can be improved without having a negative impact upon the other parameter. Value can be increased by either improving the function of the item or reducing the cost to produce the item. Value, as defined, is the function/cost. Therefore, improving the function or reducing the cost will increase the value. Basic functions must be preserved and not reduced as a consequence of pursuing value improvements. In the United States, value engineering is specifically spelled out in Public Law 104-106, which states, "Each executive agency shall establish and maintain cost-effective value engineering procedures and processes."
42. **Value Proposition:** A document that defines the benefits that will result from the implementation of a change or the use of an output as viewed by one or more of the organization's stakeholders. A value proposition can apply to an entire organization, parts thereof, or customers, or products, or services, or internal processes.

43. **Voting:** A method of selecting an item from a list of alternatives. It is a method that allows an individual to identify his or her preference from a group of alternatives.
44. **Workflow Diagram:** Visually represents the movement and transfer of resources, documents, data, and tasks through the entire work process for a given product or service. The diagram is based upon the layout of the organization or the department with which the process comes in contact.

OVERVIEW SUMMARY

There are over 1,400 different tools/methodologies related to performance improvement approaches that could be included in this book. For example, in the H. James Harrington and Kenneth Lomax book entitled *Performance Improvement Methods: Fighting the War on Waste* (McGraw-Hill, 1999), a list of 1,119 different performance improvement tools/methodology were documented. Since that time, the number of performance improvement tools and methodology has grown significantly; we estimate that a complete list would now exceed 1,500 and it continues to grow. This does not mean that there are now more than 1,500 different performance improvement tools and methodologies available to the practitioner today. What it means is that many of these "new" tools are just different names for the same old tool/methodology or very slight modification to another tool/methodology. The ones that are presented in this book are the ones that are most frequently used. These tools are the ones that are often slightly modified by individuals/consultants and given a different name in order to regenerate interest in this specific approach.

2

Affinity Diagram

DEFINITION

Affinity diagram is a technique for organizing a variety of subjective data (such as options) into categories based on the intuitive relationships among individual pieces of information. It is often used to find commonalties among concerns and ideas.

While man has been combining ideas into relational groups for years, it wasn't until 1960 when Jiro Kawakita formalized the activity coining the term *affinity diagram*. The process is sometimes referred to as the *KJ method*.

Its Uses

Affinity diagrams are used to organize the large numbers of ideas derived from brainstorming.

General

Cross-functional teams usually yield the best results from this activity.

Preparation

Start by defining the issue and gathering random ideas from a brainstorming session.

- Identify areas of concern
- Organize the ideas into the concern categories

- Prioritize the ideas in each category
- Form subgroups as ideas are sorted

The completed affinity diagram can be used as input to a cause-and-effect diagram.

Example

Start by defining the issue and gathering random ideas from a brainstorming session (Figure 2.1).

- Identify areas of concern (Figure 2.2).
- Form subgroups as ideas are sorted, and organize the ideas into the concern categories (Figure 2.3).

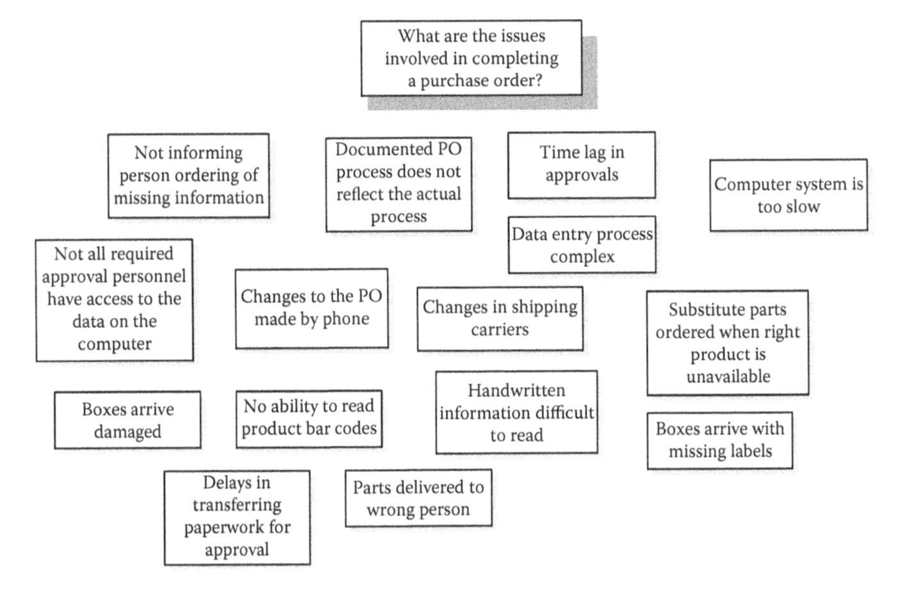

FIGURE 2.1
Brainstorming session idea generation.

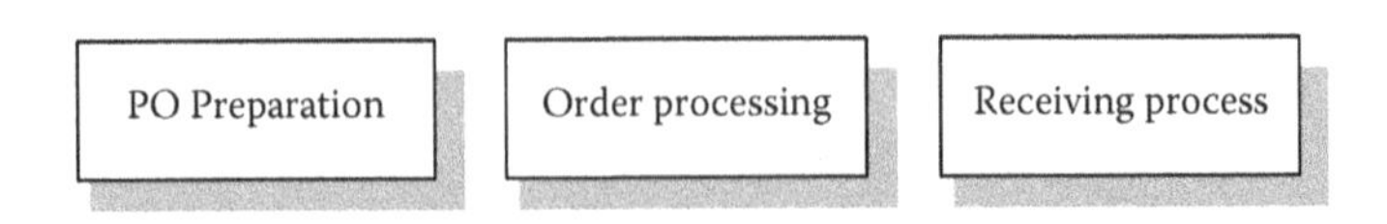

FIGURE 2.2
Areas of concern identified.

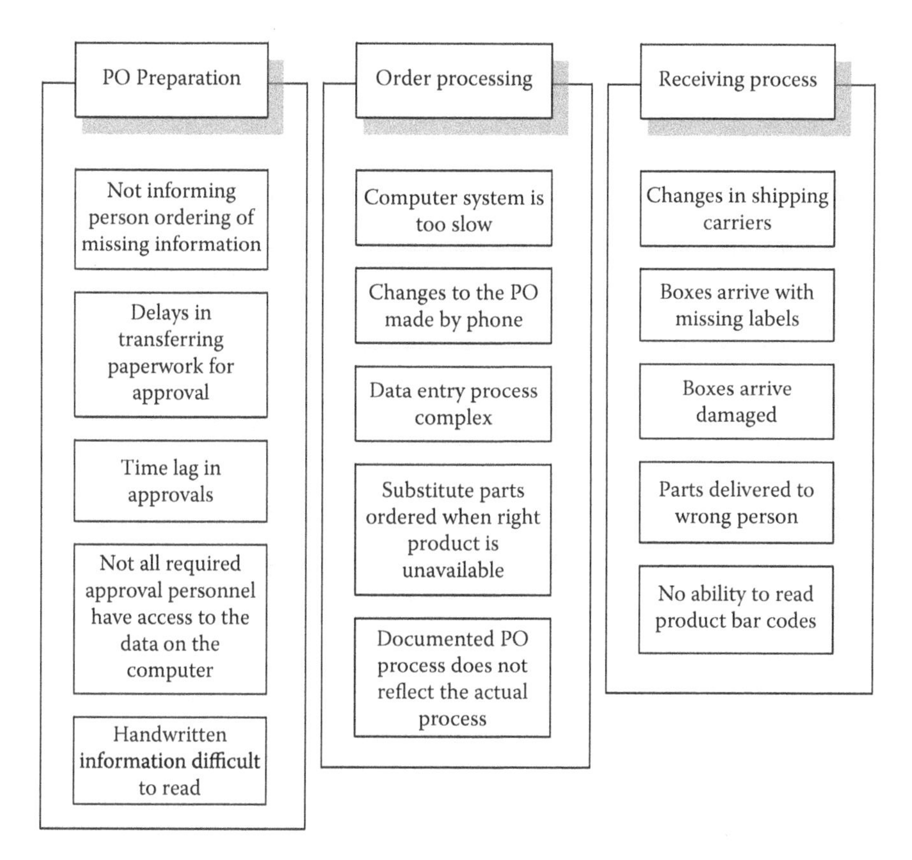

FIGURE 2.3
Concern categories.

- Separate the concern categories into actionable categories (Figure 2.4).
- Identify actionable areas (Figure 2.5).

Another example is shown in Figure 2.6.

Software

Some commercial software available includes, but is not limited to:

- Edraw Max
- SmartDraw®
- Affinity Diagram 2.1
- QI Macros™

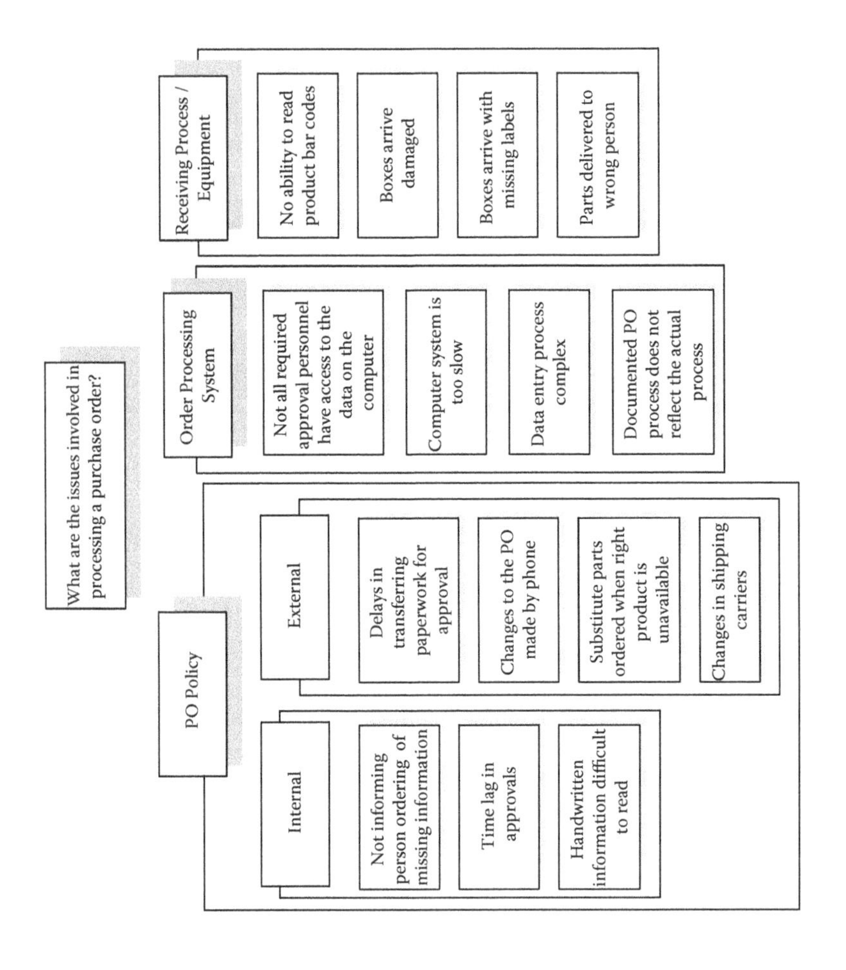

FIGURE 2.4
Concern categories separated into actionable categories.

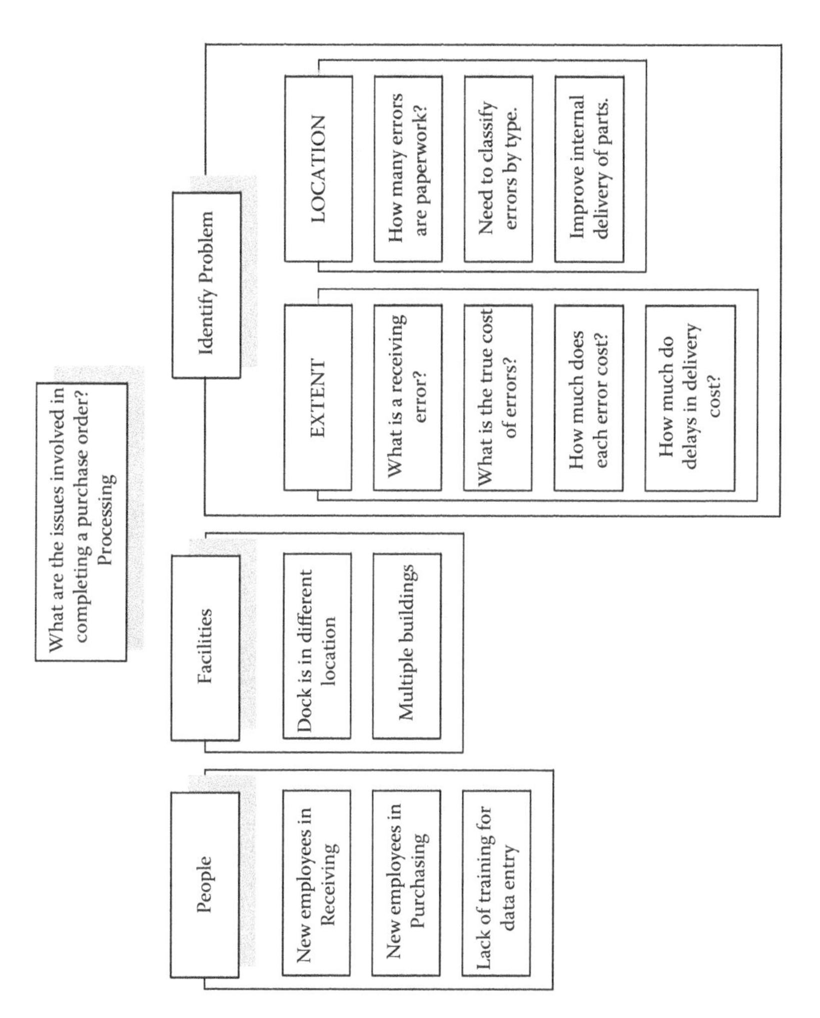

FIGURE 2.5
Final actionable categories identified.

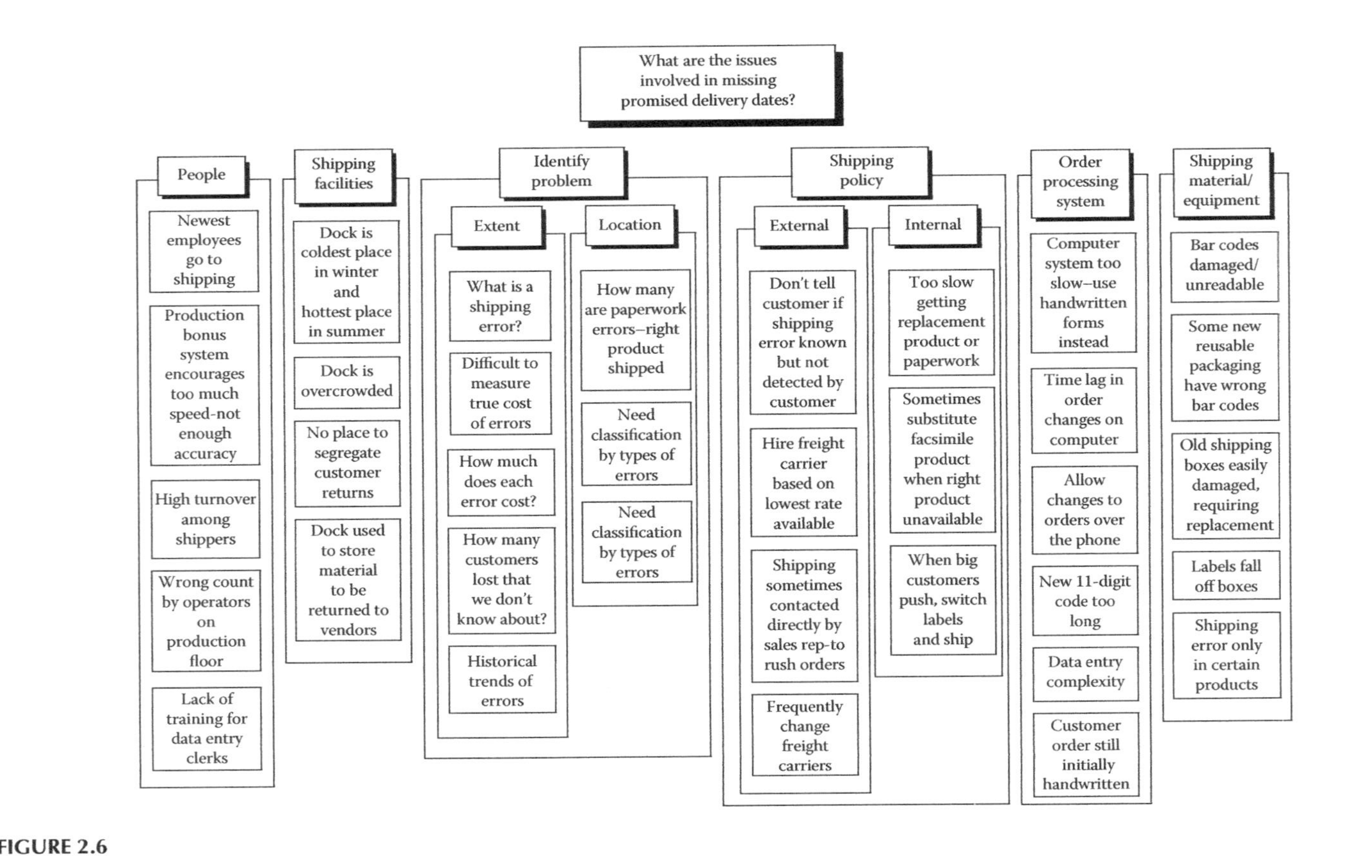

FIGURE 2.6
Another example of finalized affinity diagram.

ADDITIONAL READING

Asaka, T., and K. Ozeki, eds. 1998. *Handbook of quality tools: The Japanese approach.* Portland, OR: Productivity Press.
Brassard, M. 1989. *The memory jogger plus.* Milwaukee, WI: ASQ Quality Press.
Eiga, T., R. Futami, H. Miyawama, and Y. Nayatani. 1994. *The seven new QC tools: Practical applications for managers.* New York: Quality Resources.
King, B. 1989. *The seven management tools.* Methuen, MA: Goal/QPC.
Mizuno, S., ed. 1988. *Management for quality improvement: The 7 new QC tools*. Portland, OR: Productivity Press.

3

Brainstorming (Creative Brainstorming)

DEFINITION

Brainstorming (creative brainstorming) is a technique used by a group to quickly generate large lists of ideas, problems, or issues. The emphasis is on quantity of ideas, not quality.

Its Uses

Brainstorming can help to stimulate a wider variety of ideas. The techniques can be either structured or unstructured. Two brainstorming techniques include:

- Wildest idea
- Superheroes

General

The technique is based on the fact that a cross-functional group of people, when uninhibited, can generate numerous and creative ideas.

Preparation

- Identify and instruct a leader or facilitator
- Introduce the brainstorming session
- Welcome participants and describe the purpose of the session, roles and responsibilities, and the facilities

- Develop the ground rules for the session:
  - The facilitator establishes the ground rules with the brainstorming group members. This is important because the group members are more likely to live by ground rules they establish for themselves.
  - The facilitator may list several of the most critical ground rules. Next he or she should then solicit further suggestions from the group. When the ground rules have been agreed on, the list should be posted in clear sight of all participants. Common ground rules include:
    - Group should consist of 4 to 12 members.
    - Only one person may speak at a time.
    - No interruptions may occur until the speaker shows readiness to accept them.
    - Participant should accept all views as serious contributions.
    - Participants should speak for themselves not for other participants.
    - One idea is discussed at a time.
    - All ideas are good ones.
    - Freewheeling and building on previous ideas is encouraged.
    - No ideas are criticized.
    - Everyone is encouraged to participate.
    - All ideas are recorded.
    - All members have equal opportunity to participate.
    - Tell the group which method of participation will be used. Choose either a structured or unstructured approach:
      - A structured approach allows each person in the group to give his/her ideas as his/her turn arises in the rotation or pass until the next round. The structured method allows every member of the group an equal chance to participate.
      - An unstructured approach allows the group to give ideas as they come to mind. The unstructured method creates a more relaxed atmosphere, but also risks unequal participation from the group.
- Frame the challenge.
- Identify 6 to 12 participants including a leader and recorder.
- Take two minutes as a group to think silently.
- Call out ideas; don't allow any discussion or judgment.

Options

There are at least three general ways that brainstorming can be conducted. They include:

1. The brainstorming team agrees on the subject and takes some personal time to write down on Post-it® notes his or her comments related to the subject. They then go around from one member to the next member of the brainstorming team getting and recording the input from each member. This continues and, if an individual has nothing to comment, they just state "PASS," indicating they had no input. This continues until all the ideas have been recorded.
2. The brainstorming team agrees on the subject and takes personal time to write down its comments related to the subject on Post-it notes. While the team is recording these ideas, the facilitator records on Post-it notes some general headings related to the type of ideas that would address the problem and then posts them on the top of the idea board. When the brainstorming team has recorded its individual ideas on the Post-it notes, the members then go to the board and paste their ideas under the appropriate general headings.
3. The brainstorming team identifies a subject that warrants a brainstorming session. This subject is posted on the Internet (company or general public Internet). The news of the posting is transmitted to Communities of Interest and anyone can add their thoughts on the subject. All thoughts are recorded and made available to anyone in the Community of Interest.

Example

How to manage the brainstorming session? See the following rules:

- Ask the group members to generate a large number of ideas.
- Record all ideas where they are visible to the entire group.
- Read through the list and restate the ideas several times throughout the session.
- Don't judge any of the ideas.
- Let ideas incubate.
- After all ideas have been recorded, review and clarify them.
- Determine what action needs to be taken.

The Wildest Idea Approach

- Select a group leader.
- Define the problem or issue in a concise statement.
- Have the leader or group select a wild idea that, if feasible, would be an option to solving the problem.
- Use this idea as a starting point for the group to continue to generate ideas.
- If no practical ideas emerge, identify another wild idea and continue the process until an acceptable idea is found.

The Superheroes Approach

- Distribute descriptions of various superhero characteristics to the group.
- Instruct each group member to select one of the characteristics and assume its identity. Members may be provided with signs stating the name of the heroes they selected.
- Each group member, in turn, describes his or her character in as much detail as possible. This description should include such things as special powers, strengths, weaknesses, and habits.
- After each hero is described, group members use the information as stimuli for generating ideas that describe how the superheroes' abilities could be applied to the problem. For example, Spiderman's web might suggest a network concept for solving a problem.

Descriptions of some superheroes include:

- Superman has x-ray vision, super hearing, can fly, and is the strongest man on Earth.
- Batman and his sidekick Robin, the boy wonder, are first-rate detectives who always manage to outwit the most sinister criminals. They have at their disposal an assortment of paraphernalia, such as a Batmobile, Bat-plane, Bat-cycle, Bat-roller skates, Bat-rope, etc.
- Wonder Woman is a truly liberated woman with extraordinary strength, agility, and all-around athletic ability. She can easily overpower the most macho man. With her magic bracelets, she can deflect bullets shot at her. And, with her magic lasso, she can rope almost anything. On occasion, she flies her own invisible airplane.

- Captain America represents the ultimate in all American ideals (truth, justice, apple pie, and mom). He also has a Captain America's shield that can protect him from any harm.
- Dr. Strange tries to live up to his name. As a skilled magician and sorcerer, he can create numerous allusions. He also is able to cure sicknesses, control people and situations, and change one thing into something else.

Software

Some commercial software available includes, but is not limited to:

- MindGenius® MindMap
- SmartDraw® MindMap
- QI Macros™

4

Cause-and-Effect Diagram

DEFINITION

A cause-and-effect diagram is a visual presentation of possible causes of a specific problem or condition and their relationship to each other. The effect is listed on the right-hand side and the causes take the shape of a fish bone. It is for this reason it is sometimes called a Fishbone diagram.

Its Uses

These diagrams are used for product design, identifying causes of undesired results and quality defect prevention. They identify potential factors contributing to the cause of a result from a process. The causes are divided into major areas that can contribute to the result under investigation. The areas, or categories, most commonly used include:

- **People:** Anyone involved with the process.
- **Methods:** Procedures and work instructions used in producing the result.
- **Machines:** All tools and equipment used to accomplish the job.
- **Materials:** Raw materials, parts, etc., required to produce the final product.
- **Measurements:** Variables that control the process and the results of the process.
- **Environment:** Location, time, temperature.

General

Cause-and-effect diagrams help support risk analysis studies, such as failure modes effect analysis and root cause analysis. The cause-and-effect diagram was one of the basic tools taught by Dr. Ishikawa in the quality circle movement in Japan. Due to Dr. Kaoru Ishikawa's focus on the importance of using the tool, it became widely used throughout the world and, as a result, it is sometimes referred to as Ishikawa diagrams.

Preparation

- Identify the problem or effect and place it in a box on the right-hand side of the available writing space.
- Draw a long horizontal line with an arrow pointing to the box.
- Identify the major categories of causes.
- Identify causes and group them into major categories.
- Identify additional causes using brainstorming.
- Write the detailed causes in clusters around the major categories they influence.
- List the series of steps in the process you wish to analyze.
- Identify all the causes that contribute to the process.
- Write each cause around the appropriate step on the diagram.
- Evaluate each cause:
  - Brainstorm both positive and negative effects
  - Identify any interrelationships
  - Brainstorm proposed solutions
- Using the information, choose the solution to use.

Example

See Figure 4.1.

Software

Some commercial software available includes, but is not limited to:

- Edraw Max
- SmartDraw®
- Affinity Diagram
- QI macros™

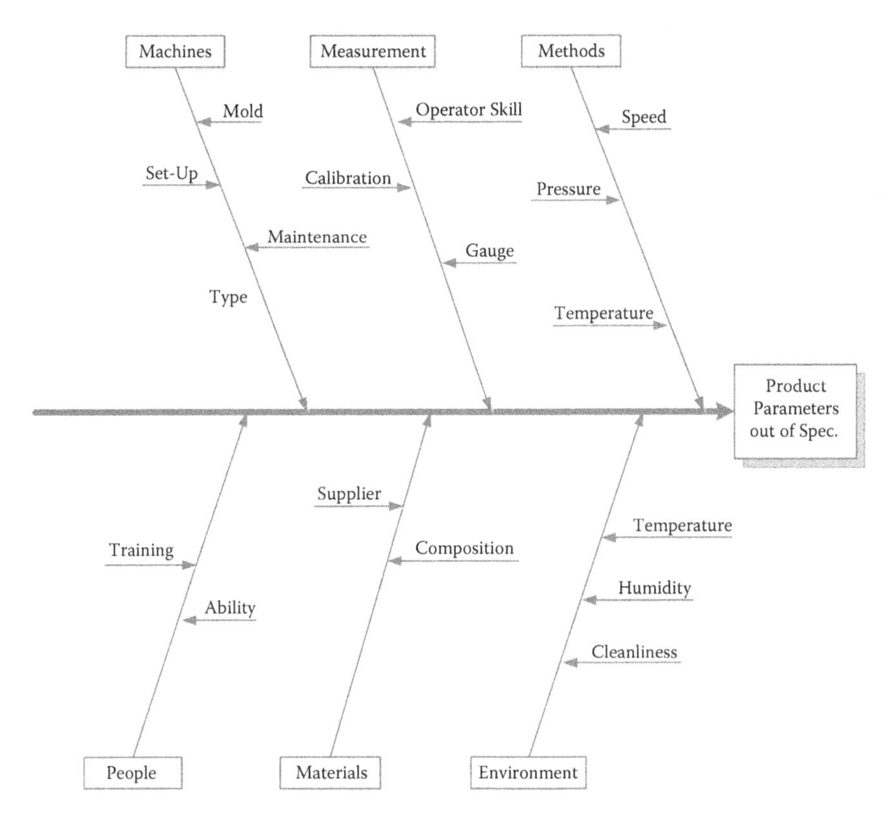

FIGURE 4.1
Completed cause-and-effect diagram.

5

Check Sheet

DEFINITION

A check sheet is a simple form on which data are recorded in a uniform manner. The forms are used to minimize the risk of errors and to facilitate the organized collection and analysis of data.

Its Uses

Recording check sheets are used to collect different types of data while location check sheets are usually graphics on which data are indicated. Checklists are used to verify that tasks have been completed.

General

Check sheets ensure that all relevant data are collected and all required activities are completed for a given task. An example is an airline pilot's checklist prior to takeoff and landing.

Preparation

- Identify all the data to be collected.
- Determine additional information to be collected (who, when, where, etc.).
- Design the check sheet supporting the information to be collected.
- Check sheets should be as simple and clear as possible.
- Go through the check sheet form prior to using it.
- Train appropriate employees in the use of the check sheets.
- Monitor the data collection process and identify areas for improvement.
- Modify the collection process as appropriate.

Example

Location check sheet (Figure 5.1).
Recording check sheet (Figure 5.2).
Operation check sheet (Figure 5.3).

Software

Some commercial software available includes, but is not limited to:

- Airwire

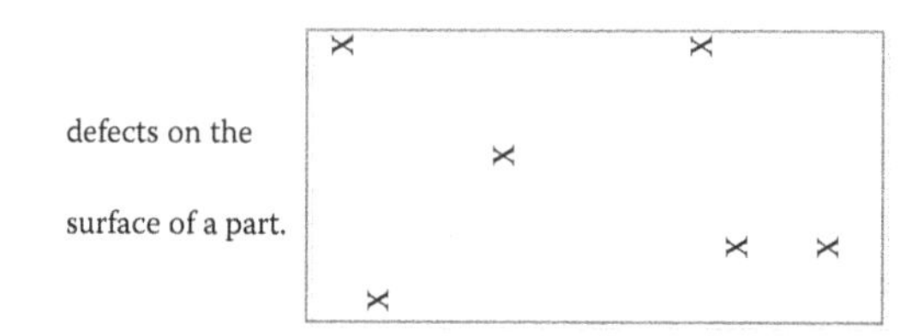
defects on the
surface of a part.

FIGURE 5.1
Location check sheet.

Week Ending / Defects	1/6	1/13	1/20	1/27	2/3	2/10	2/17	2/24	3/1	3/8	Total
Wrong Component(s)	I	I	III	I	IIII	II	I		IIII	IIIII IIIII II	29
Wrong Base Material			III		IIII	III	III				13
Misreads	I	I	IIII		IIII		I		III	IIIII II	21
Sequence Count Off		III	I		IIII	II	I		III	IIIII	19
Parts Count Off			I							I	2
Total	2	5	12	1	16	7	6	0	10	25	84

FIGURE 5.2
Recording check sheet.

Machine Set-Up Check Sheet			
Machine ID:	Operator:	Date:	Time:
Setting	Target	Actual Reading	Comments
Spindle Speed	2100–2300 rpm		
Temperature	120–130°F		
Angle	23–24 deg.		
Pressure	48–50 psi		

FIGURE 5.3
Operation check sheet.

6

Commitment Building

DEFINITION

Commitment is a promise to give or do something, to be loyal to someone or something. It is the act of pledging or engaging oneself in an obligation or a promise to be engaged, or becoming involved in a given activity to achieve a given result.

Its Uses

It is used in organizational change to improve how workers feel about their jobs and their contribution to the final goals and outcomes.

General

Organizational commitment seeks to predict work variables that will affect the outcome of any given activity. It seeks to mitigate and eliminate elements that reduce an individual's commitment to the company and project.

Preparation

Building commitment starts with the identification of the elements involved in organizational change and methodology. Executives must develop a firm commitment to the change project. This is usually accomplished through executive briefings identifying and supporting the desired future state model.

There are three stages of commitment and eight levels of change:

1. Preparation
   a. Contact (unawareness to awareness)
   b. Awareness of change (confusion to process)

2. Acceptance
   a. Understand the change (negative perception to positive perception)
   b. Positive perception (decision to support the installation)
3. Commitment
   a. Installation
   b. Adoption (change accepted in most areas)
   c. Institutionalization (change accepted in all areas)
   d. Internalization (full commitment)

The critical lessons associated with successfully building commitment during change include:

1. Commitment is expensive; do not order it if you cannot pay for it.
2. Commitment strategies must be developed.
3. Building commitment is a developmental process.
4. Either build commitment or prepare for the failure.
5. Human reaction to change is a function of intellectual and emotional response cycles.
6. Recognize the power and responsibility of learning how to manage change.

Example

Establish contacts with participants:

1. Accurate, concise, and clear information must be communicated to those affected by the change.
2. Establishing the intended awareness may require several contact efforts.
3. The desired outcomes of establishing contact include:
   a. Unawareness: Reduce the probability that adequate preparation for commitment will occur
   b. Awareness: Advances the preparation process

Develop an awareness of the change:

1. Information concerning the change must continue to flow.
2. Find different ways of communicating the information to all concerned.
3. Everyone must get the same information.

4. The desired outcomes for developing awareness of the change include:
   a. Confusion: Diminishes adequate preparation
   b. Understanding: Advances the process toward acceptance

Cultivate understanding of the impending change:

1. With awareness and understanding, individuals can make judgments about the change.
2. With understanding the change and its implications, people will still have questions and concerns.
   a. Negative perception: Decreases the level of support and provides the first opportunity for resistance
   b. Positive perception: Increases support for the acceptance of the change

Foster a positive perception of the change:

1. With a positive perception of the change, the individual will choose whether or not to support implementation of the change.
2. Perception of change agents and targets is important to the implementation decision.
3. The desired outcome for producing the positive perception of change could be:
   a. A decision not to support installation or
   b. A formal decision to initiate and utilize the change.

Implement/install the change:

1. When installation occurs, the "commitment threshold" has been reached.
2. During implementation, if problems become too expensive or cumbersome, pacifism would increase and may result in persons no longer supporting the change.
3. Some pessimism concerning the change is unavoidable; however, confidence is increased as a result of resolving such problems.
4. The desired outcome for the implementation/installation phase include:
   a. The change is adopted after initial utilization.
   b. Adoption: Promotes long-term use of the change.

Adopt the change:

1. After a successful trial period, adoption occurs.
2. The degree of the change must be assessed.
3. The desired outcomes from adoption include:
   a. The change is aborted after extensive utilization or
   b. Institutionalization: Represents the highest level of organizational commitment possible.

Encourage institutionalization of the change:

1. With institutionalizing, full change has been achieved.
2. A considerable amount of time usually is needed to move from installation through adoption to institutionalization.

Promote internalization of the change:

1. Maximum support is achieved when participants are driven by an internal motivation that reflects their own beliefs and wants as well as those of the organization.
2. In order to internalize the change, those involved must truly comprehend the perceived cost of the change.

7

Consensus Building

DEFINITION

Consensus building is a technique to obtain the commitment of all team members to move in a particular direction.

Its Uses

Consensus building is used when a team is unable to unanimously define a single solution.

General

Consensus building does not establish a solution that all members agree on, but instead establishes the solution with which all members can live.

Preparation

Establish ground rules with all participants:

1. Establish a set of basic ground rules with each participant.
2. All team members are given a chance to state their views.
3. Members listen to and consider each other's ideas.
4. Alternatives are seriously considered by all members.
5. Disagreements are opportunities for gaining insights and broadening the range of opinions.
6. Members initiate discussions of the process when the group's tasks are not being accomplished.
7. Every member of the group is willing to give the final decision a trial for reasonable period of time and to participate in its implementation.

Example

Soldering defects have increased and production is down. Bring the employees together to discuss the problem and find a solution.

- Create a list of solution options:
    - Resolder the defectives on employees' own time
    - Retrain
    - Post pictures of good and bad solder joints
    - Change the assembly process:
        - Instead of baking individual parts, bake multiple parts
        - Process in an assembly line manner instead of all operations by one person
- Attempt to find a solution that is agreeable to all participants:
    - Discuss each suggestion
    - List each by most agreeable to least agreeable
- Evaluate the remaining choices:
    - Eliminate the bottom choices
- Review each solution from the perspective of each team member:
    - Rank the remaining solutions and discuss the best solution or solutions
- Select the candidate solutions:
    - The final solution was:
        - Have each person do all soldering steps.
    - Bake multiple parts instead of one at a time.

This activity resulted in the reduction and elimination of bad solder joints and an increase in productivity of 18 percent.

8

Consequence Management

DEFINITION

Consequence management is a formal process of understanding the institutional structures that reflect deeply held values and beliefs in the organization and then utilizing those structures (e.g., compensation or training opportunities) to influence desired behavior.

Its Uses

Using these management levers increases the probability of desired behavior (i.e., that which is directed toward achieving the future state) and to inhibit behavior that supports the current state.

General

Consequence management can be used to pull people toward the future state, or help push people from the current state.

Preparation

How do you use consequence management to manage change? It can be used to manage change by pulling people toward the future state in one or more of the following ways:

- Decrease the rewards for maintaining the current state. For example, current state processes may value the volume of widgets produced and have no corresponding offset for poor quality. If quality is desired, then a consequence management strategy may be to use existing rewards, punishments, and effort to provide an incentive for an operator to switch from volume production to quality production.

- Increase punishment for current state behavior. This would be a more active next step by making it harder for an operator to produce volume by only counting those widgets that pass a quality test as complete widgets.
- Support the rewards for moving to the new standard. This step may involve offering training in how to be one's own quality inspector.
- Concurrently, along with the push away from the current state, design a variety of incentives, rewards, and recognition programs to pull and support movement toward the future state:
  - Some of the tactics employed in consequence management include making it more difficult to maintain the present state and making it easier to act consistent with the goals and objectives of the future state.
  - To encourage movement away from the current state, the rewards for maintaining the current state need to be decreased, the punishments increased, and the amount of effort needed to maintain current state needs to be increased. At the same time, to encourage behaviors consistent with the future state, rewards need to be increased, punishments decreased, and effort level reduced with respect to behaviors and actions consistent with the future state.
- Determine how sponsor effectiveness will be measured.
- Identify the performance parameters that the sponsor is responsible for and how to measure the performance against the schedule.
- Identify existing positive reinforcements:
  - Determine what already exists in the organization (positive reinforcement) that can be used to reward behaviors and actions that are consistent with the change effort. These positive reinforcements do not necessarily have to be financial; in fact, often more beneficial are nonmonetary rewards, such as a pat on the back, pins, public mentions, hats, T-shirts, lunches, employee of the month, etc.
- Identify existing negative reinforcements:
  - Determine what negative reinforcements can be used for people not supporting the change or leading nonproductive attacks on change objectives. Examples may include not getting a good

review, good versus bad assignments, counseling/reassignment/ redeployment, or discipline.

- Create a penalty for negative behaviors:
  - Create a penalty to be applied to those who resist beyond a reasonable amount of time. Design strategies to make it harder to maintain the current state and easier to act consistent with the future state behaviors. Examples might include desirable assignments made on the basis of future state performance rather than seniority or volume standards that now count for only one-half their former value.
- Generate a detailed listing of reinforcements and penalties:
  - Generate a detailed listing of these positive and negative reinforcements and effort level strategies from the sponsor's perspective.
- Involve sponsors in determining appropriate reinforcements:
  - Meet with the initiating sponsors and/or sustaining sponsors and decide which reinforcements they are comfortable using to encourage behaviors consistent with the change objectives and discourage those behaviors that interfere with the change objectives.
- Identify the consequence management actions:
  - Identify actions for consequence management; how and when to use positive and negative reinforcements. Provide the sponsor with action plans that implement the consequences.
- Use consequence management to accomplish change:
  - Two important tasks must be accomplished if change is to happen. The first task is to motivate change, or to overcome the natural resistance to change that emerges. This involves getting individuals motivated in ways that are consistent with the immediate change goals and with long-range organizational effectiveness. The second major task is managing change. We can think of organizational changes in terms of transitions from a current state through a transition state and into a future state. An important issue in any change effort is how to effectively manage the transition from current state to future state. Use positive and negative reinforcements (consequence management) to encourage transition.
  - Rewards and punishments can play a major role in making the transition a successful one. One reason individuals resist change is that often change is perceived as a threat to their job,

pay, and success. The effective use of consequence management can motivate individuals to act in ways consistent with transition and the new desired state.

- Assess the ability and willingness of the sponsors to implement consequence management:
  - If the initiating sponsor or sustaining sponsor is unable to utilize the prescribed consequence, provide training in the skills, etc., that they need. On the other hand, if the initiating sponsor or sustaining sponsor is unwilling to utilize consequence management in conjunction with the change objectives, it may be necessary to go to a higher level of sponsorship in the organization.
- Obtain sponsor commitment to consequence management plan:
  - Confirm that the initiating sponsor will commit to employing consequence management techniques where appropriate in order to facilitate implementation efforts. Identify sponsor effectiveness measurements and determine how sponsor effectiveness will be measured.

Example

An example was given in this section under *Preparation*.

9

Control Chart

DEFINITION

A control chart is a graphic representation that monitors changes that occur within a process by detecting variation that is inherent in the process and separating it from variation that is changing the process (special causes).

Its Uses

Control charts represent individual points or distributions that are plotted continuously over time and monitor the performance of a processor activity. The purpose of the control chart is to monitor and distinguish the difference between normal causes of variation and special causes of variation in a processor activity. This is done by setting up statistical confidence limits to indicate when a point is most likely not expected.

General

Variables Control Chart

Variables Control Charts are created from quantitative and monitor the trends of a specific characteristic. Examples include pressure, temperature, weight, etc.

There are three types of variable charts:

- Chart of individuals (charts individual values) and a running range chart.
- A mean in range chart (charts the means of readings taken at a given time and their range).
- When there are more than 10 samples in each subgroup, instead of using a range chart, use a standard deviation chart.

Attributes Control Chart

Attributes Control Charts represent qualitative data and count the number of occurrences of defects. Examples include breaks, scratches, failures, etc.

There are four types of attribute control charts used for statistical analysis:

- P-chart for the percentage of nonconforming units or also percent yields
- Np-chart for the number of nonconforming units
- C-chart for the number of nonconformities
- U-chart for the number of nonconformities per-unit

Preparation

Variables Control Chart

- Collecting the data:
  - Determine the type of data to be measured and whether they will be individual measurements or subgroups from which the charted values will be calculated.
    - A subgroup is a set of measurements taken at the same time:
    - There should be sufficient time between subgroups to determine variation over time.
  - Define the frequency of collecting a subgroup data and how it is to be collected. The data will represent the variations within a process at a point in time in the variation or process over time.
- Calculate the mean values in ranges:
  - Calculate the mean for each subgroup in the range within each subgroup:
    - The range for each subgroup is the difference being the largest and smallest values of that subgroup. If individual readings are to be taken, then the range is calculated by the difference between each successive reading.
  - The average range is calculated from all of the range values.
- Calculate the control limits for the mean chart:
  - The control limits are calculated from all of the data collected over time. (It is recommended that a minimum of 20 to 30 readings should be used before the control limits are calculated).

TABLE 9.1

Control Limit Factors

(Subgroup Size)	A_2	d_2	D_3	D_4
2	1.880	1.128	0.0	3.268
3	1.023	1.693	0.0	2.574
4	0.729	2.059	0.0	2.282
5	0.577	2.326	0.0	2.114
6	0.483	2.534	0.0	2.004
7	0.419	2.704	0.076	1.924
8	0.373	2.847	0.136	1.864
9	0.337	2.970	0.184	1.816
10	0.308	3.078	0.223	1.777

- The control limits for the X-Bar charts are calculated by multiplying X-Bar by the appropriated A2 factors based on the number of samples in each subgroup (Table 9.1).

  The control limits for the X-Bar charts for samples less than or equal to 10 in the subgroup are calculated as follows:

  Upper control limit: X-Bar + A2(R-Bar)
  Lower control limit: X-Bar − A2(R-Bar)

  Example:
  For a sample of subgroup size = 5 and 20 readings:

  X-Bar = 10.66
  Upper control limit
  10.66 + (0.577 × 1.59) = 11.58
  Lower control limit
  10.66 − (0.577 × 1.59) = 9.74

- Calculate the control limits for the range chart:
  - The control limits for the range chart are calculated by multiplying the mean of the range by a factor in the following table corresponding with the number of observations in each subgroup. e.g.:

    R-Bar = 1.59 (calculated from 20 means of samples with 5 readings each)
    The control limits are calculated as follows:

    Upper control limit for R:
    D4(R-Bar) = 2.114 × 1.59 = 3.35
    Lower control limit for R:
    D3(R-Bar) = 0 × 1.59 = 0

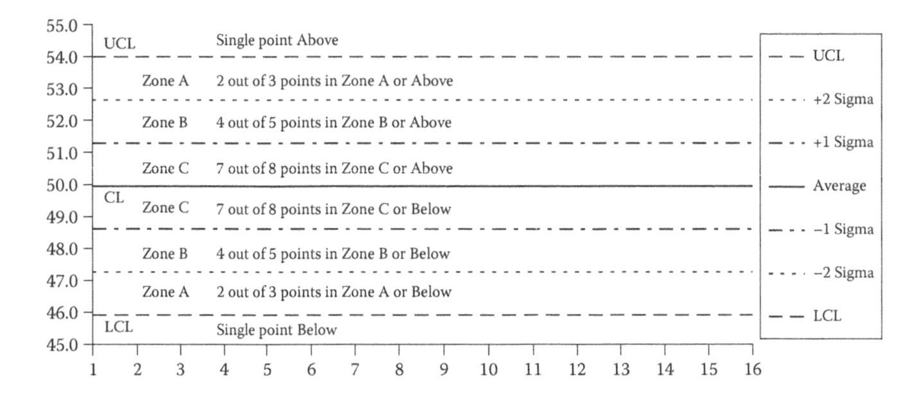

FIGURE 9.1
Pattern rules requiring attention.

Control Chart Constants

The upper and lower control limits are equivalent to the plus and −3 sigma levels for a population. When the chart is constructed, we divide the range between the mean in the upper control limit by 3 and add the 1 into sigma lines. We do the same for the difference between the lower control limit in the mean.

Rules for Identifying Special Causes of Variation

There are five general rules for patterns of data that indicate that a potential special cause of variation has occurred (Figure 9.1). They include:

1. One point above the upper control limit or below the lower control limit
2. Two points out of three above the 2 sigma line or below the lower 2 sigma line
3. Four points out of five above the 1 sigma line or below the 1 sigma line
4. Seven out of eight points in a row below the mean or above the mean
5. Seven points in a row continually rising or falling

There are other rules that also apply and special rules can be made for specific types of data.

Example

See Figure 9.2 and Figure 9.3.

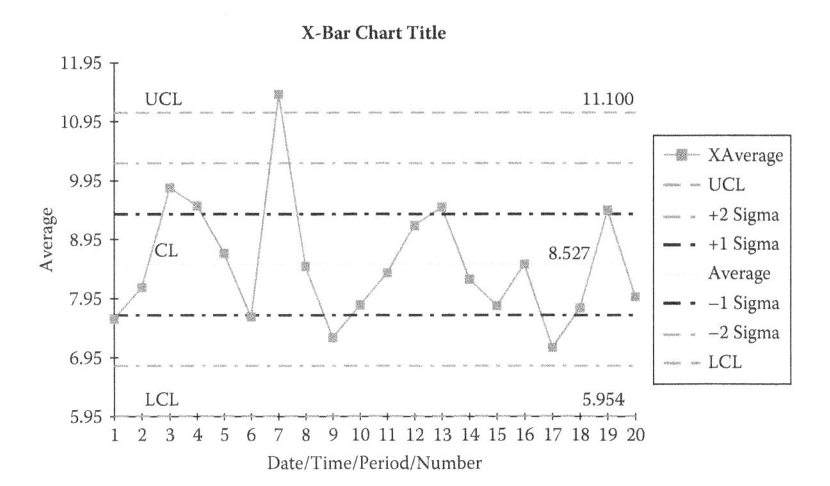

FIGURE 9.2
X-Bar chart title.

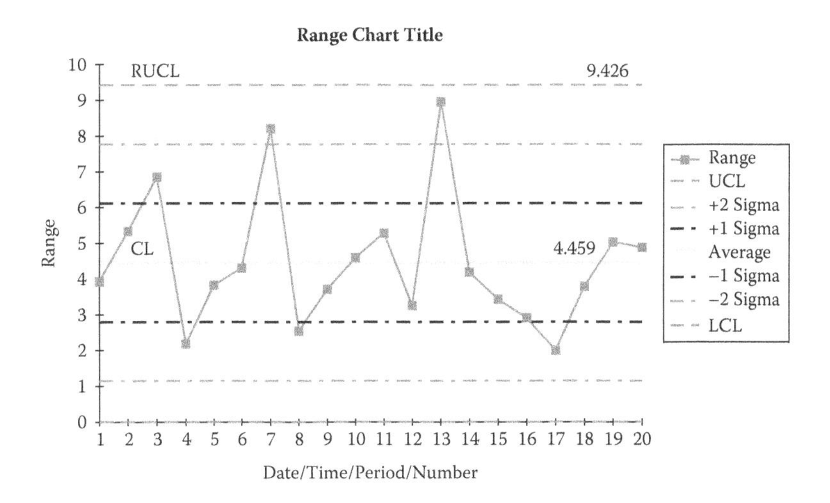

FIGURE 9.3
Range chart title.

Software

Some commercial software available includes, but is not limited to:

- Minitab®
- JMP®
- QI Macros™
- Sigma XL®

ADDITIONAL READING

Box, G., and A. Luceno. 1997. *Statistical control by monitoring and feedback adjustment.* Milwaukee, WI: ASQ Quality Press.

Burr, I. W. 1976. *Statistical quality control methods.* New York: Marcel Dekker.

Deming, W. E. 1986. *Out of the crisis.* Cambridge, MA: MIT Center for Advanced Engineering Study.

Feigenbaum, V. 1961. *Total quality control: Engineering and management.* New York: McGraw-Hill.

Grant, E. L., and R. S. Leavenworth. 1988. *Statistical quality control,* 6th ed. Milwaukee, WI: ASQ Quality Press.

Harrington, H. J., G. Hoffherr, and R. Reid. 1998. *Statistical analysis simplified: The easy-to-understand guide to SPC data analysis.* New York: McGraw-Hill.

Ishikawa, K. 1982. *Guide to quality control,* 2nd ed. Tokyo: Asian Productivity Organization.

Shewhart, W. A. 1931. *Economics control of quality manufactured product.* New York: D. Van Nostrand Company. (Reprinted by ASQC.)

10

Cost–Time Chart

DEFINITION

A cost–time chart is a date-and-cost line chart that tracks a process changing cost over time. (Also referred to as a date-and-price chart, and similar to a Gantt chart.)

Its Uses

These diagrams are used to display broad process trends and predict future costs. Retail and wholesale businesses that replenish stock continually use this sort of graph to monitor their own expenses. These graphs also are used to track the price of the business product and to record and predict revenue. As with all charts that track data over time, the independent variable must go on the chart's x-axis. Cost, the dependent variable, goes on the chart's y-axis.

General

The chart can be used to track any process over time relative to its cost. This chart is similar to a "value stream" chart used in Lean.

Preparation

Create the cost–time chart:

- Draw the axes for an X-Y graph. Show cycle time on the X-axis and cumulative cost on the Y-axis. Plot both axes in terms of cumulative values.

- Identify each activity in the process by drawing vertical lines corresponding to the cumulative cycle time of that part of the process and all previous activities. List the activities in chronological order, from left to right, on the graph. Label each segment on the graph with the appropriate event name.
- Draw the cumulative line to represent the cost per unit cycle time of each segment. This line represents the contribution to total cost of each activity or task within each segment.
- Analyze the plotted data:
    - The cost between the start and end of each activity represents the cost and time for that activity. The accumulation of all of the times and costs represent the overall time and cost for the completed project.
- Identify activities for further investigation.

Identify activities in the chart that are candidates for improvement. Look for items with:

- High level of resources
- Long required time
- Long wait time
- High cost

Example

See Figure 10.1.

Software

Some commercial software available includes, but is not limited to:

- RFFlow

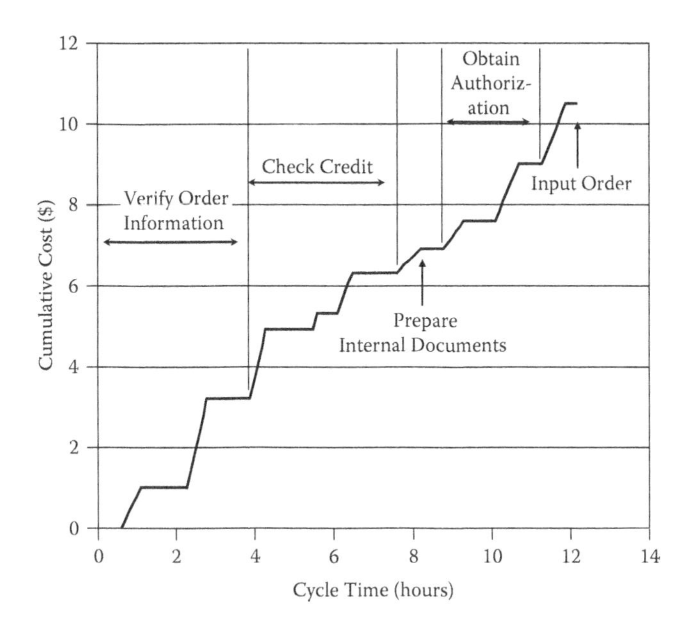

FIGURE 10.1
Sample cost time chart.

11

Data Gathering by Document Review

DEFINITION

Data gathering is a technique to quickly collect information that currently exists within an organization.

Its Uses

Data gathering is used for collecting data with minimal interruption and use of resources.

General

Document review can be used to assess the current extent of documentation and record keeping within a company, which can provide a good baseline of its level of organization and control. Document review can be used, in conjunction with other data gathering techniques, to collect information for further analysis. This is similar to the audit of an ISO standard.

Preparation

Gathering Data by Document Review

Preparing to gather data includes the following:

- Identify the processes for which information is needed and decide what processes will be the subject of document review. Determine the informational needs related to the selected processes.

- List all available documents related to the processes. Assess their usefulness in terms of their completeness, clarity, and whether they are up to date.
- Some documents that may be considered for document review:
    - Interoffice memoranda
    - Meeting minutes
    - Accounting records
    - Standard operating procedures
    - Job descriptions
    - Completed transaction forms
    - Manual and computerized files
    - Manual and computerized reports
- In addition, information may be obtained from the following sources:
    - Existing flowcharts and flow diagrams
    - Software documentation
    - Training manuals
- Gather documents that are most appropriate for review based on their usefulness and their match to the informational needs:
    - Gather information from the documents
    - Review documents to gather required information
    - Review additional documents beyond the original list if necessary
- Organize the findings. Once the necessary data are gathered, the findings should be organized for future reference and for use with other techniques.

Example

An ISO registration or surveillance audit evaluates how well the documentation reflects how the processes are being done and how well they meet the intent of the ISO standard being audited.

12

Data Gathering by Interview

DEFINITION

Data gathering by interview is the act of using the interviewing process to collect data.

Its Uses

Data gathering by interview is used for collecting data from people who are actually doing the work. This also captures attitudes.

General

Interviewing and interpersonal skills are critical for gathering accurate and detailed information required by the problem-solving process. Information gathered during an interview can include, but is not limited to:

- Goals and objectives
- Business functions
- Problems and concerns
- Critical assumptions
- Critical success factors
- Information needs
- Organizational structure
- Entities
- Attributes
- Relationships
- Events
- Processes
- External agents

There are two types of interviews: single interviews, involving one or sometimes two interviewees, and group interviews in which three or more participants are interviewed simultaneously. The two interview types follow the same basic procedure:

- Prepare for the interview
- Conduct the interview
- Summarize the interview
- Confirm the interview findings

This technique can be used with any other technique that requires information from company representatives.

Preparation

Preparing for the Interview

- Select an appropriate number of interviewers and identify who will lead the interview.
- Ensure that the interviewer has the necessary communication, listening, and facilitation skills to guide the interview successfully. The interviewer also should have some analytical skills related to the relevant project phase.
- Clearly identify the objectives of the intended interview. This makes it easier to correctly identify the people to be interviewed and to keep the interview properly focused.

Identify Candidates for the Interviews

- List possible interviewees based upon the objectives of the interview.
- When an appropriate interview candidate is identified, obtain the consent of his or her manager before approaching that person.
- When possible, find out the duties and responsibilities of the interviewee (perhaps from his or her manager).
- If a specific individual cannot be identified, ask the appropriate manager to recommend one. When possible, arrange for the manager to make the introductions.
- Do not include an interviewee's manager or supervisor in the same interview. Doing so can inhibit the discussion.

Arrange the Time and Place for the Interview

- Select a mutually convenient time and place where interruptions will be minimal.
- Provide sufficient notice for interviewees to plan to attend.
- Arrange for any visual aid requirements.

Prepare and Distribute Appropriate Briefing Materials to the Interviewees

- Identify areas the interview is to cover and, if appropriate, prepare a checklist of specific questions.
- Identify the topics that the interview will cover. Prepare a checklist of specific questions for each.
- Prepare an interview guide to identify the purpose of the interview. The interview guide should include:
  - Background: Briefly describe the project and the context of the interview.
  - Purpose: List and describe the objectives of the interview.
  - Topics to cover: List the topics and the most significant questions that the interviewer should address.
- Prepare any other briefing materials needed to support an effective interview.
- Issue the interview guide and any other appropriate briefing materials to the interviewees well in advance of the interview, to allow sufficient preparation time.
- Contact user interviewees individually before the meeting, if practical, to confirm their understanding of the meeting's objectives and any briefing material provided.

Conducting the Interview

General interviewing tips:

- Be patient, listen to each viewpoint, explain the problem. Follow the interviewee's preferred sequence of material.
- Start with a top-down approach. It helps to define the grounds to be covered during the interview, which, in turn, helps you judge the progress. Before leaving a point, check your understanding with the appropriate closed questions.

- Use visual aids, such as whiteboards and flip charts, as much as possible. The whiteboard is a useful way of involving the interviewee and recording findings.
- Be fully up to date on the current state of the analysis. Review all models, even if they are in the process of being developed and refined. This helps you focus on incomplete or weak areas of the developing analysis.
- When preparing for a program of interviews, briefly review any available material to acquire some familiarity with basic terms in business activities. This can save time during the interview and help establish your credibility.
- Establish a relaxed tone for the interview.
- Introduce the participants and interviewers.
- Confirm the intended length of the interview, its purpose, and its context within the project. Put the group at ease and invite open discussion.
- Follow the structure given in the interview guide.
- Address the topics outlined, but be flexible about the order, as needed.
- Ensure that digressions do not take excessive time away from central issues.
- Make note of ideas to come back to later in the discussion. Remember to address them.
- Do not try to cover too much ground at one time. If necessary, make an appointment to continue another time.
- Ask the appropriate interview questions.
- Use open questions, which avoid "yes" and "no" answers.
- Allow time for considered responses. Do not cut off thoughtful silences.
- Rephrase a question if it appears to have been misunderstood.
- Ask clarifying questions to elicit further details and to separate opinions from facts.
- Ask for examples and for copies of any current documentation mentioned. These can illustrate difficult or new concepts.

Note-Taking

During an interview, it is usually necessary to take notes. These notes are typically unstructured and hard to follow for anyone but the interviewer. Therefore, because interviewing is usually followed by focusing sessions, it is advisable to restructure the notes into a format suitable for review. This format can vary based on the objectives of the interview.

Example

Questioning Technique

Central to all investigating is the questioning technique. Avoid questions that are too general (e.g., "What do you do in your job?"). Such questions have no focus; the response could lead anywhere and could quickly become difficult to control. Instead ask, "What are your five to eight main responsibilities?" Always have a purpose in mind when asking a question. Questions can be classified as follows:

- Open: Difficult to answer with a "yes" or "no." These questions usually begin with the 5 Ws and an H. (This type of question drives the main body of the interview and keeps conversation flowing):
  - Who
  - What
  - Why
  - When
  - Where
  - How
- Closed: Invites a "yes" or "no" answer. This type of question should be used only to gain confirmation or denial of a point. For example, "If I understand you correctly, you are responsible for this activity?" Used too often, the closed question leads to a very stilted conversation.
- Leading: Tells the interviewee what you want to hear. For example, "So, you, in fact, must be responsible for this activity?" Such a question often leads to a "yes" or "no" answer and causes problems similar to those experienced with closed questions.

13

Data Gathering by Samples

DEFINITION

A sample is a representation from a population that allows the observer to predict the actual distribution of the population. There is rarely enough time or resources to measure the entire population; a representative sample of the population will yield a model for the entire population.

Populations are sometimes obvious, like a batch of parts, but sometimes sampling must be done over space, time, or a combination of dimensions. In other cases, performance of a system becomes the population, such as gambling probabilities, e.g., the performance of a pair of dice, a slot machine, keno balls, or a roulette wheel, over time to determine if there is recognizable pattern or "nonrandom" distribution.

Its Uses

Sampling is used to reduce the time and effort and cost related to collecting information. It must be used correctly in order to get satisfactory results, e.g., using too small a sample size or if the sample is not representative of the total population. The process of data gathering by samples is important to ensure that the data are being collected from the appropriate population and in the same manner. If data are collected from different populations, valid conclusions cannot be drawn and the evaluation will be inconclusive.

General

There are two types of sampling: probability sampling and nonprobability sampling. In probability sampling, every member of the population has a chance of being selected in the sample and that distribution can be accurately determined. This allows for an unbiased estimate of the population distribution by identifying members of the population by their probability of being selected.

In nonprobability sampling, it is possible for some of the members of the population to have no chance of being selected. It usually involves sample selection based on assumptions about the population. Because the members of the population are nonrandom, this technique does not allow us to estimate sampling errors. Therefore, the relationship between the sample and the population is limited and it becomes difficult to extrapolate from the sample to the population.

There are many ways to determine sample size. Well-known approaches include tables, formulas, and pour function charts.

The following lists the variety of sampling methods:

- Random Sampling:

  In this case all samples taken from the population have an equal probability of being selected. This also is true of any multiple samples taken from the population. This technique minimizes bias and simplifies analysis. This also allows us to use the variance within the sample to represent the variance of the population.
- Systematic Sampling:

  Systematic sampling relies on a selection or ordering scheme and selecting elements at regular intervals from that ordered selection. This usually starts with a random selection of the first sample and then taking every k^{th} member until the final sample is obtained.
- Stratified Sampling:

  When a population is made up of a number of distinct categories, the sampling plan can be divided into these categories or "strata." Each stratum is sampled as an independent subpopulation from which individual members are randomly selected. Some advantages of stratified sampling include:
  - Inferences can be drawn about specific subgroups that may have been lost in the overall population.
  - Can lead to more efficient statistical estimates.
  - Sometimes preexisting strata within a population are more readily available.
  - Because each stratum is evaluated independently, different sampling approaches can be used.

Conditions to using a stratified sampling approach include:
  - Variability within strata are minimized.
  - Variability between strata are maximized.
  - The selected stratum is correlated with the dependent variable of interest.
- Probability Proportional-To-Size Sampling:
  In some cases, the distribution by size becomes an important characteristic and, therefore, selecting samples by size can significantly reduce the number of samples collected. An example would be the type of particles that pass through a filter.
- Cluster Sampling:
  Sampling by groups can be more cost-effective, such as geography, time periods, etc.
- Quota Sampling:
  In quota sampling, the population is segmented into mutually exclusive subgroups similar to stratified sampling. Samples are then drawn from each segment and the segments are analyzed separately and then compared to each other.
- Accidental Sampling:
  This type of sampling also is known as opportunity, convenience, or grab sampling. This is a nonprobability sampling drawn from easily obtained members of a population.
- Line-Intercept Sampling:
  Selecting a sample based on a secondary condition exists that may be totally independent of the sample, but exists at the time or place where the samples are collected.
- Replacement versus without Replacement Sampling:
  In "without replacement sampling," no member of the population can be selected more than once in the same sample. In "replacement sampling," a member of the population may be selected more than once in the same sample.

Sampling and Data Collection

Appropriate data collection schemes involve:

- Defining and following a sampling process
- Maintaining the time and location of each sample
- Labeling contextual elements and comments
- Identifying nonresponses

Preparation

When gathering data through sampling, it is important to collect information from a sample that is representative of the entire population. In general, a more accurate representation will be attained with larger sample sizes. As a guideline, the sample size should be at least 30; however, the required sample size can be calculated based on an intended confidence level and precision. Assuming a normally distributed population, a representative sample size can be statistically determined using a specific confidence level (e.g., 90 percent confidence in the sample results), precision (e.g., +/–5 of the stated result), and the following formula:

$$N = z^2\sigma^2/B^2$$

where

- N = Required sample size
- z = 1.645 for a 90 percent confidence level (1.964 for a 95 percent confidence level 2.575 for a 99 percent confidence level)
- σ = – Standard deviation of the population
- B = Intended precision bound (margin of error)

Example for a double sided test:

How many samples need to be taken to have a 95% confidence that the sample mean is within 1 unit of the population mean μ. Assume that the population $\sigma = 6.95$ units.

$$N = z^2\sigma^2/B^2$$

$$N = (1.964 \times 6.95)^2/1 = (13.62)^2 = 185.55$$

Rounding we get N = 186 samples

Example

A product is made from a continuous flow process machine running 24 hours a day, 365 days a year. In order to monitor the process, samples need to be taken and measured. From historical data, it is determined that the process is stable over time with some unpredictable machine failures that do not affect the quality of the product.

It was decided to take samples of the product every two hours and perform the quality measurements. If the sample does not pass any of the critical tests, the existing product is put on hold until additional testing can be performed.

Software

Some commercial software available includes, but is not limited to:

- Minitab®
- JMP®

14

Data Gathering by Surveys

DEFINITION

Surveys are used to measure characteristics of a population relative to such things as behavior, awareness of programs, attitudes or opinions, and needs. Repeated surveys can give valuable information about trends, such as in evaluating government activities.

Its Uses

Surveys are frequently used to project how people will feel about a particular product, situation, or individual candidate. It is used to predict future results based on how individuals feel related to the item being surveyed. The process of data gathering by surveys is important to ensure that the data are being collected from the appropriate population and in the same manner. If data are collected from different populations, valid conclusions cannot be drawn and the evaluation will be inconclusive.

General

Surveys gather information from a sample of the population and, therefore, the sample size is dependent on the purpose of the survey. In any survey, the sample must be objectively chosen such that each member of the population will have a nonzero chance of selection. Without this condition, the data can't be reliably projected to the population.

The major disadvantage of a survey is lack of control over the data collected. Factors to consider when determining which type of survey to use include:

- Can the data be collected accurately and effectively?
- Is the topic sensitive?
- Is the topic complex?
- Are the respondents qualified to supply the required information?
- Will respondents be willing to supply the information?
- How valid will the responses be and how will you validate the responses?

Other considerations before performing a survey include:

- Accuracy: What is the acceptable level of error?
- Frequency: Is the survey to be repeated and how often?
- Ethical consideration: Will the data be handled confidentially?

Survey Process

The survey process is complex and steps are not necessarily sequential in nature. Development of the survey is critical and may require more than one set of evaluations and modifications in order to produce a satisfactory structure. Key steps in producing a survey include:

- Planning and designing
  - What is the purpose and objectives? What question(s) will be addressed and answered?
  - What will be the sample selection method?
  - What will be the collection methodology?
  - Question design is critical and may need to address the same area from different points of view in order to draw valid conclusions.
- Testing and modifying
  - Run a pilot survey
  - Analyze the pilot results
  - Modify questions according to the analysis
  - Repeat the pilot as necessary
- Conduct the survey
  - Finalize the questions and collection process
  - Select the sample
  - Train the data collectors
  - Conduct the survey

- Processing and analysis
  - Define data entry and format
  - Enter the data for analysis
  - Analyze the data:
    - Calculate population estimates
    - Determine standard errors
  - Draw conclusions
  - Prepare the report

Data Collection Method

The most commonly used survey methods for collecting quantitative data include:

- Telephone interviews
- In-person interviews
- Self-completion questionnaires:
  - Mail
  - Email
  - Web-based
  - Short message service (SMS)

Self-Completion Surveys

These are usually the least expensive especially for reaching a large sample. Respondents can take their time to answer the questions and refer to reference material where needed. The major disadvantage is size of the nonresponders and of those that do respond, a sample that does not represent the population of interest.

In-Person Interviews

Face-to-face surveys usually produce more data to be gathered than self-completion surveys and can use more complex inquiries.

Telephone Surveys

These are generally cheaper and quicker than in-person surveys and can be done quickly when timely results are desired. While more calls can be made in a given time, the nonresponse can be high.

Computer-Assisted Telephone Interviewing (CATI)

This type of telephone survey can be conducted much faster and answers are entered directly into the database allowing for much faster data analysis.

Sources of Error

Errors fall into two groups: sampling and nonsampling errors.

- Sampling Errors: This type of error occurs when the sample is not representative of the entire population. The factors affecting sampling errors include:
    - Sample size: A larger sample size may reduce the sampling error.
    - Variations in the population.
    - Sample design: Occurs when the probability of sample selection is not known.
- Nonsampling Errors: Table 14.1 is a summary of nonsampling errors.

Bias and Accuracy

Refusals, noncontact, and language differences are factors that produce nonresponses in all surveys. Bias also can come from inadequate sampling plans. For example, if you are interested in surveying to see how people will vote in an election, but do not get a representative sample from all party affiliations, it will lead to a biased result.

Example

See Figure 14.1 (Tire survey results from TireRack.com.)

Software

Some commercial software available includes, but is not limited to:

- Minitab®
- JMP®

TABLE 14.1

Nonsampling Errors

Source of Error	Examples	Strategies to Minimize Error
Planning and interpretation	Poor definition of terms, questions and population identification and question(s) to be answered.	Precisely define all concepts, terms, and populations between data users and survey designers.
Sample selection	Too small of a sample size, biased sample selection, etc.	Validate the questions, remove any duplicates, and add any clarifying questions.
Survey methods	Inappropriate method (e.g., mail survey for a very complicated topic).	Carefully choose the appropriate survey method.
Questionnaire	Misleading or ambiguous questions, questions structured to evoke a single answer.	Use understandable language, clear questions, and validate through testing.
Interviewers	Leading, making assumptions, misunderstanding, or misreporting answers.	Train interviewers in how to ask questions and to not embellish the respondents' answers.
Respondents	Uncooperative respondents.	Ensure confidentiality; use well-trained, impartial interviewers and probing techniques; if mail-based, use a well-written introductory letter.
Processing	Errors in data entry, coding, or editing.	Adequately train and supervise data processing and validate data periodically.
Estimation	Incorrect weighting, errors in calculation of estimates.	Ensure that skilled statisticians undertake estimation.

				TIRE PERFORMANCE RATINGS								
				WET		DRY			COMFORT			
	Rank Within Category	% vs. Best In Category	Would Buy Again	Hydroplaning Resistance	Wet Traction	Cornering Stability	Dry Traction	Steering Response	Ride Comfort	Noise Comfort	Treadwear	Total Miles Reported
Bridgestone Potenza RE-11	1	100%	8.9	7.9	8.2	9.4	9.5	9.2	8.2	8.2	7.9	1,889,260
Dunlop Direzza ZII	2	98%	8.5	8	8.5	9.3	9.5	9.2	7.7	6.8	8.2	176,990
Yokohama ADVAN Neova AD08	3	96%	8.4	7.8	8	9.2	9.4	9.4	7.8	7.6	7.1	440,866
Hankook Ventus R-S3	4	90%	8.7	6.5	6.8	9.1	9.2	9	7.4	7.1	7.2	843,933
BFGoodrich g-Force Rival	5	90%	8	6.3	7.1	8.7	9.1	8.9	7.2	7.1	7.6	111,245
Kumho Ecsta XS	6	90%	7.8	6.3	6.7	8.9	9.2	8.7	7.7	7.4	7	1,022,771
Bridgestone Potenza RE070	7	85%	7.2	6.3	6.7	8.9	9.1	8.9	6.2	5.7	6.7	2,802,473
Goodyear Eagle F1 Supercar G: 2	8	81%	4.9	5.6	5	8.6	8.1	8.8	7.2	7	5.3	160,425

Color Key

Info based on consumers completing Tire Rack's online survey. "10" is the highest. See key below.

Tire Performance Ratings:

Superior 8.6-10 Excellent 6.6-8.5 Good 4.6-6.5 Fair 2.6-4.5 Unacceptable 0-2.5

Would buy again?

Definitely 8.6-10 Probably 6.6-8.5 Possibly 4.6-6.5 Probably Not 2.6-4.5 Definitely Not 0-2.5

FIGURE 14.1
Tire survey results. (From www.tirerack.com/tires/surveyresults/surveydisplay.jsp?type=EP. With permission.)

15

Data Stratification

DEFINITION

Data stratification is a technique used to help identify the underlying causes of variation within a population of data.

Its Uses

The technique is used in conjunction with other data gathering techniques and tools. It may be useful when you discover unexpected results during data analysis.

General

The process breaks down large categories into smaller ones for data collection and analysis. This allows you to recognize when a subcategory significantly contributes to the variation within a category. Looking only at the combined data masks the contribution of the individual elements.

It is most beneficial when it occurs before data collection. This prevents the timely task of recollecting data for each subpopulation. When collecting data, try to identify areas where subpopulation data may be useful in determining causes of variation. Based on these areas, design the data collection device accordingly.

Preparation

- Identify data for which analysis might benefit from stratification.
  Look for potential abnormalities in the distribution of data.

- Use data stratification as a diagnostic tool …
  Data stratification results in a graphical representation of the sources of variation in a data population. Stratified data can help to determine root causes or areas for process improvements.
- Identify factors that might contribute to distribution abnormalities.
  Conduct a brainstorming session to identify factors affecting the distribution of the data.
- Stratify the data.
  Arrange the data into subpopulations. Determine which variables to collect data on at the subpopulation level, e.g., by shift, machine, machine operator, type of defects, etc. Sometimes this step can occur through the simple separation of previously collected data. Other times, the data must be recollected for each subpopulation.
- Plot the new or newly stratified data.
  Plot the new data for each subpopulation selected, using the appropriate graphic technique or tool, such as histograms, Pareto analysis, or any other presentation method.
- Compare the plotted data with that for the entire population.
  Look for differences between the plot for each subpopulation and the entire population. If the subpopulation plots are similar to that of the entire population, identify new variables that might be causing unusual variation in the data. If the individual subpopulation plots are different from the plot of the entire population, then the variable that distinguishes the subpopulations is at least one cause of abnormal distributions for the entire group.

Example

See Figure 15.1.

Software

Some commercial software available includes, but is not limited to:

- Minitab®
- JMP®
- QI Macros™
- Sigma XL®

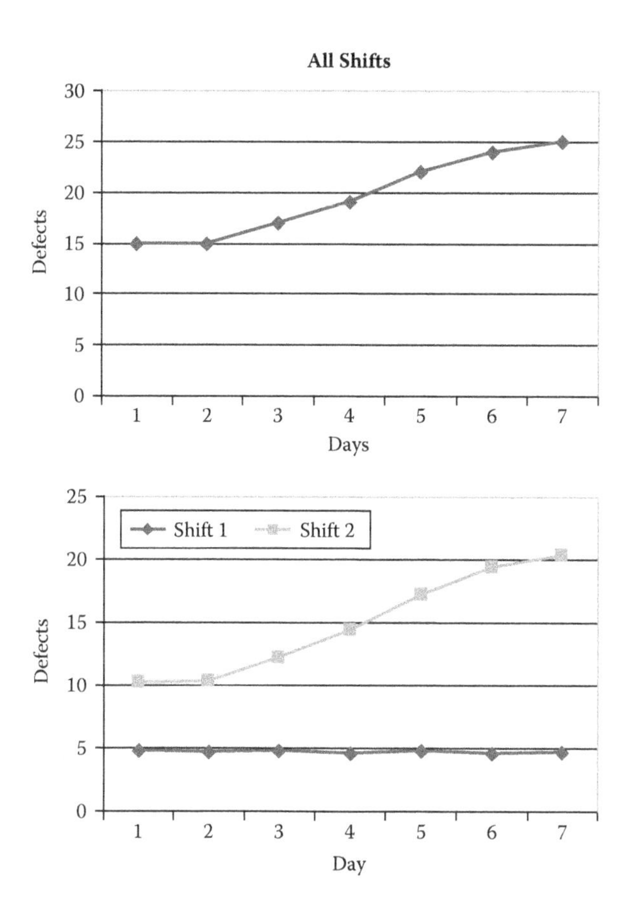

FIGURE 15.1
Sample data stratification graph.

16

Delphi Narrowing Technique

DEFINITION

The Delphi narrowing technique is a tool that eliminates the need for face-to-face interaction while it enables achieving group consensus through the use of a prioritization scheme.

Its Uses

The Delphi narrowing technique is useful when group participants are not in the same physical proximity, anonymity is desired, personalities are likely to cause disruptive conflict, or various levels of participants may cause intimidation in a face-to-face setting.

General

The Delphi narrowing technique does not allow for the following:

- Opportunities for creative interaction
- Independent generation of solutions and prioritization of choices
- Prevent "groupthink."

The Delphi technique facilitates a group of individuals to address a complex problem. The technique is a systematic solicitation of opinions on a given topic utilizing questionnaires and summaries on the feedback. The key aspect of this process is that the respondents do not get together and their responses are anonymous.

Preparation

- Agree on the criteria for narrowing the combined list of possible solutions.
- Ask participants individually to list possible solutions.
- Ask the participants to narrow the list of solutions.
- Supplement to Delphi narrowing technique with other techniques when appropriate.
- Ask each person to prioritize the selected solutions.
- Create a new master list.
- Use the new list for further evaluation.
- Repeat the narrowing process if necessary.

Example

Using the Delphi Narrowing Technique

1. Agree on the criteria for narrowing the combined list of possible solutions:
   a. Possible criteria for narrowing the combined lists include:
      i. Sample focusing criteria include the solutions' effects on:
         (1) Product/service quality
         (2) Time
         (3) Cost
         (4) Resources
   b. Have members challenge the decision criteria, if appropriate, until consensus is reached.
2. Ask participants, as individuals, to list possible solutions.

   Provide a problem statement and ask each individual to develop a list of possible solutions.
3. Ask the participants to narrow the list of solutions.
   a. Collect all solution lists and combine the results to create the master list. Distribute the master list to each member.
   b. State the agreed decision criteria and instruct each individual to independently choose a minimum number of master list solutions for further consideration.

      General guidelines for the number of choices to allow each participant are listed below.

      40–60 Solutions: 6 Choices

      25–40 Solutions: 5 Choices

15–25 Solutions: 4 Choices
10–15 Solutions: 3 Choices
5–10 Solutions: 2 Choices

For example, if a list of 12 solutions is developed, ask each team member to narrow the list down by choosing two solutions.

4. Supplement Delphi narrowing technique with other techniques when appropriate:
    a. It may be useful to develop a master list using an interactive technique, such as brainstorming:
        i. Develop the master list by whatever method is most appropriate.
        ii. Ask the group to follow the rest of the Delphi narrowing technique to independently prioritize preferred subsets of solutions.
5. Ask each person to prioritize the selected solutions.

    Prioritize his or her selected solutions and assign a number value to each solution according to its priority. A high priority item should receive the highest numerical value. For 6 solutions, the highest priority solution would be given a 6 and the lowest a 1.
6. Create a new master list:
    a. Consolidate all the participants' chosen solutions into one combined master list.
    b. Tally the number of responses given for each item on the list.
    c. Circle the solutions that received the strongest indications for further consideration.
    d. To select the top alternatives, circle the ones receiving:
        i. The highest total value
        ii. The highest number of mentions
        iii. A top priority by any participant
7. Use the new list for further evaluation.

    Submit the narrowed list of alternatives for further discussion, investigation, data collection, etc.
8. Repeat the narrowing process if necessary.

    If the group has not reached adequate consensus on what solution to implement, then repeat the cycle of asking participants to select and prioritize a smaller number of possible solutions.

17

Employee Involvement

DEFINITION

Employee involvement (EI) is a technique for unleashing human potential in organizations and involving people in the change process.

Its Uses

Involving people in the change process substantially increases the likelihood of their support for change objectives. This can be accomplished by delegating power and decision making to lower levels in the organization and, by using concepts like shared visions of the future, to engage all employees so that people develop a sense of pride, self-respect, and responsibility.

General

Employee involvement is the sum of many parts:

- It is a partnership.
- It is a process in which groups work together to create a climate where everyone can achieve job satisfaction by directing their ingenuity, imagination, and creativity toward achieving change objectives.
- It is a means of providing employees the opportunity to identify and resolve problems related to the change.

Employee involvement is a positive process. It contributes to:

- job satisfaction
- giving people an opportunity to more fully utilize their experience, talents, and ideas

- giving employees an important role in identifying and solving problems
- fostering open communication between all involved parties
- improving relationships in the workplace
- increasing the likelihood of successful change implementation

When employees are involved in the implementation of a change, they tend to give their support and commitment to the change initiative and enable implementation of the change objectives.

Preparation

For employee involvement to work, sponsors must play a critical role. Sponsors need to be dedicated to the concept and willing to devote the necessary time and resources. They also need to adopt a participative management approach that features open communication, trust, cooperation, listening, and a willingness to involve the employees to a greater degree in the change process.

Example

1. Using employee involvement:
   a. Obtain sponsor involvement and support.
      Get the sponsor's strong commitment to the change initiative and up-front public support of the project.
   b. Establish an executive sponsor and transition team.
      Sponsorship will cascade down through the organization by forming executive sponsor and transition teams. The essential functions of these teams are to:
      i. identify likely projects and determine their objectives;
      ii. develop procedures and approaches;
      iii. implement pilot projects;
      iv. provide on-going support for projects; and
      v. resolve problems if and when they arise.

      The transition team normally consists of representatives from all parts of the organization. It should not be larger than 6 to 10 members.

      All executive sponsor team and transition team meetings should follow a prescribed format, including an agreed-upon agenda and regular, defined time frames. The executive sponsor team may, in its entirety or as a subset, function as the transition team, depending on the size of the organization. The transition team is responsible for performing an analysis of the organization and for developing plans for the change implementation.

2. Analyze the organization:
   a. Identify factors that might facilitate or impede the change objective.

      The transition team may do this by conducting a change readiness assessment to determine the enablers and barriers of the change at a high level in the organization.
   b. The following factors might be examined when diagnosing the organization's readiness for change:
      i. Supervisory practices
      ii. Workforce practices
      iii. Task and role characteristics
      iv. Work group
      v. Problem-solving process
      vi. Job satisfaction
      vii. Involvement/commitment
      viii. Performance
      ix. Recognition
      x. Resistance
      xi. Culture
3. Develop a framework for the change project.

   The transition team develops the framework for the change project. Factors to consider include, but are not limited to, the level of commitment required to achieve the change objectives and the details of the change initiative. In the beginning, keep the project details quite flexible.
4. Prepare the organization.

   The transition team informs and prepares the organization for the change. Preparation includes providing employees with an understanding, knowledge, and confidence in employee involvement and change management. Some activities to accomplish this include awareness presentations and other types of communications, and orientation sessions.

5. Establish change agent teams.

   Set up change agent teams after awareness training and other preparatory efforts are completed. The change agent teams are responsible for developing an approach for implementing the enablers and barriers analysis with target groups of employees in the organization. The analysis involves assessing the enablers and barriers that exist toward the change from the target groups. This cascade of assessment activities fulfills two purposes:
   a. First, it is conducted to reach the transition team and change agent teams as targets to communicate the future state context down and feedback perceptions up the organization.
   b. Second, it serves to gather information from the targets to inform planning decisions.

      Note: Target groups need to include at least 15 to 30 percent of the target population.
6. Successful change will cascade and spread through the organization.

   The degree of employee involvement should continue to increase as the change project progresses and the organization moves closer to reaching the future state objectives.
7. Evaluate and modify the framework for implementing change.

   Always pilot the changes in a small area of the organization. From this, experience analysis and planning activities may need modification before implementing the change on a large scale. After the plans are modified, change teams can begin to work with other areas of the organization, empowering teams to achieve the change objectives.

18

Facilitated Session

DEFINITION

A facilitated session is a meeting in which the leader (facilitator) guides the discussions through a series of steps designed to arrive at a consensus that is acceptable to all participants. A facilitated session helps the participants to define and support mutual goals and objectives.

Its Uses

The purpose of facilitated sessions is to attain a high level of consensus based on factual knowledge in compressed time frames to meet a predetermined set of deliverables.

General

The idea of facilitation has been around since the 1970s and was originally developed by IBM Canada as part of its Joint Application Development (JAD) approach to better application design. An important part of this methodology is the selection of an individual (facilitator) who enables groups and organizations to work more effectively. A trained independent facilitator will have a wide range of skills and techniques within his/her repertoire and will likely be knowledgeable and adept in such areas as group development, advanced communication and listening, group dynamics, conflict resolution, creativity, and innovation.

Four guiding principles for facilitators include:

1. Remain neutral at all times
2. Focus on process, not content
3. Position the group as the expert
4. Never do anything that the group can do for itself

Preparation

- Prepare a checklist of activities and areas of concern
- Look over the facilities the day before the workshop sessions begin
- Establish the roles of facilitated session participants:
  - Session facilitator
  - Scribe
  - Project sponsor
  - User participants

A facilitator is an individual trained in how to run a meeting efficiently and effectively. However, he/she may or may not actually be running the meeting. In many cases, the facilitator acts as an unbiased individual who is focusing on the way the meeting is managed to minimize the time required to come up with the most beneficial answer. Facilitators are extremely confident at helping the group to arrive at a current consensus related to what action should be taken. The facilitator should follow these standards:

1. Ensure that the purpose of the meeting is well defined.
2. Ensure that the correct participants will attend the meeting.
3. Ensure that the agenda is well organized and distributed ahead of time.
4. Ensure that the facility is properly laid out and the correct equipment is provided.
5. Ensure that the discussion during the meeting stays on track.
6. Encourage all participants to contribute to the discussion.
7. Remain neutral during the meeting and ensure that alternative ideas are fairly discussed.
8. Ensure that the minutes of the meeting reflect what went on during the meeting and the decisions that were made.
9. Ensure that the minutes of the meeting are distributed shortly after the meeting is completed.

Example

Introduce the facilitated session:

1. Welcome participants and describe:
   a. the process;
   b. roles and responsibilities; and
   c. the facilities.

2. Introduce the participants.
3. Describe the session objectives.
4. Give management the opportunity to make opening remarks.
5. Agree on the ground rules for the session:
   a. Only one person may speak at a time.
   b. No interruptions until the speaker accepts them.
   c. All views should be accepted as serious contributions by the participants.
   d. Each participant should speak for themselves, not for the others.
   e. One idea is discussed at a time.

Manage the agenda:

1. Be flexible and willing to change methods for achieving the same results.
2. Where a particular case study is used, updated data should be provided as frequently as possible.

Facilitation models:

1. One individual dominates the group:
   a. Regain control without embarrassing individual
   b. Involve other members of the group
   c. Use the ground rules ("Don't speak again until someone else has spoken.")
   d. Use a technique to get other participants involved:
      i. Brainstorming
      ii. Nominal group
2. Low rate of participation or one or more participants has not spoken:
   a. Utilize active listening techniques:
      i. Establish and maintain eye contact with the speaker
      ii. Acknowledge what the speaker says
      iii. Use nonverbal signals
      iv. Ask for clarification and expansion
   b. Con people who haven't spoken directly.
   c. Go around the room and ask each individual to voice an opinion:
      i. Brainstorming
      ii. Nominal group

3. A participant offers an idea that is ignored by the group:
   a. Bring attention to the idea that was offered.
   b. Write down or otherwise record the idea.
   c. Ask for further discussion of the idea.
   d. Thank the participant for the idea.
4. A participant offers an idea that is ridiculed or attacked:
   a. The ground rules for participants include no putdowns.
   b. Every idea gets fair consideration.
   c. Write down or otherwise record the idea.
   d. Thank the participant for the idea.
5. Participant has a hard time expressing an idea or that idea is badly stated:
   a. Use active listening to encourage the speaker.
   b. If necessary, ask questions to clarify.
   c. Restate the idea in more concise form.
   d. Thank the participant for being willing to contribute.
6. One or more participants ask you for your opinion:
   a. Acknowledge the question and state that it is more important to hear from the group first.
   b. Restate the question and ask the group to respond to it.

Managing the group's focus:

1. A participant becomes emotional:
   a. Remain calm.
   b. Make eye contact with the individual.
   c. Acknowledge the emotion.
   d. Identify the issue behind the emotion.
   e. Take steps to resolve the issue.
   f. Post it on the open issue list.
   g. Ask if it is possible to move on.
   h. If necessary, take a break and speak personally to the individual.
   i. Refocus the group.
2. The group gets off topic or individuals interrupt each other or there are constant interruptions (cell phones, etc.):
   a. Comment on what you observe.
   b. Remind participants of the ground rules.
   c. Refocus the group.

3. Two or more participants get stuck in a conflict of opinions:
   a. Comment on what you observe.
   b. Remind participants of the purpose of the meeting.
   c. Express confidence that agreement can be reached.
   d. Identify/list items participants agree upon.
   e. Identify/list items participants disagree upon.
   f. Resolve disagreements.
   g. Refocus the group.
4. Two or more participants continually engage in private conversations:
   a. Establish direct eye contact with the participants involved.
   b. Do not embarrass the participants involved.
   c. If behavior persists and it is disruptive to the group, remind the group as a whole of the ground rules.
   d. If this behavior still persists, have the group take a break and speak to the participants personally.
5. There is movement toward action without adequate discussion or their agreements with no designated action or one member is moving the group toward a single point of view:
   a. Use your voice:
      i. Lower volume
      ii. Drop tone
   b. Comment on what you observe.
   c. Get the group to generate ideas by:
      i. going around the room and ask each person to voice an opinion;
      ii. conducting a brainstorming session;
      iii. using nominal group technique; and/or
      iv. dividing the group to discuss pros and cons.
   d. Write down all ideas and issues.
   e. Cannot move on until key issues have been resolved.
6. The group members seem bored or restless or there is difficulty in resolving an agenda topic or participants are groping for ideas:
   a. Use your voice:
      i. Raise the volume
      ii. Raise your tone
      iii. Speak faster
   b. Comment on what you observe.
   c. Increase your movement in front of the group.

   d. Bring the present agenda topic to closure by:
      i. suggesting a possible solution;
      ii. delegating the solution to someone; and/or
      iii. posting the issue on a "to be discussed" list and coming back to it later.
   e. Ask the group if they are ready to move on.
   f. If necessary, take a break.

Ensure participant involvement, refine the agenda as needed, conclude the session.

1. Stick to the agenda and close the session when the major topics have been covered or time is running out.
   a. Review:
      i. All issues have been recorded and read by the scribe.
      ii. Responsibilities and commitment dates are assigned to each action item.
      iii. The work covered is reviewed.
   b. Evaluation:
      i. Participants evaluate the effectiveness of the workshop.
      ii. Workshop effectiveness is evaluated by the facilitator and management.

Analyzing and documenting facilitated sessions, assess the workshop outputs.

1. Compile the information generated in the workshop session.
2. Information from the workshop may consist of:
   a. Notes
   b. Flipchart sheets
   c. Worksheets
   d. Issues lists
3. Formalize all of the output into a summary report.
4. Evaluate the outputs against the objectives of the workshop.
5. Identify any shortcomings.

Develop a facilitated session feedback package, distribute the feedback package.

Monitor and resolve open issues, verify facilitated sessions results.

1. Determine the best way to review the feedback package:
   a. Individual meeting
   b. Group or focusing session
   c. Written memo

Schedule participant review, conduct participant review of feedback package.

1. Focus primarily on the accuracy and completeness of the work products.
2. Assign responsibility for follow-up on new issues.
3. Reiterate any tasks required to incorporate the review comments.

19

Flowchart

DEFINITION

Flowcharting is a method of graphically describing an existing or proposed process by using simple symbols, lines, and words to pictorially display the sequence of activities. Flowcharts are used to understand, analyze, and communicate the activities that make up major processes throughout an organization. They are essential tools used in process redesign, process reengineering, Six Sigma, and ISO documentation.

Its Uses

A flowchart is used to analyze systems and processes and to understand how they function. They show the route that an item will follow as it moves through the process. It is used to train employees, estimate cost to process the item and to improve the effectiveness and efficiency of the process.

General

There are many different types of flowcharts. It is important that the practitioner select the best one for the specific application (Table 19.1).

Preparation

Flowcharts have application in almost all parts of the problem-solving process. They are useful for identifying problems, defining measurement

TABLE 19.1

Seven Types of Flowcharts

Type	Description
1. Process blocks: (Block Diagram)	Document "what" is done, to illustrate a high-level flow of operations.
2. Process charts	Document "how" by breaking down a process under study into activities chronologically.
3. Procedure charts	Document the detailed flow of activities in the process.
4. Functional flowcharts	Document the process, emphasizing responsibilities and interaction between departments.
5. Geographical flowcharts	Document the physical movement of people and/or materials.
6. Paperwork flowchart	Document the detailed flow of paperwork forms within a process.
7. Information flowcharts	Document office procedures (manual and automated) using standard IS0 17BM symbols.

points for data collection, idea generation, and idea selection. The usefulness of flowcharts has been recognized to such an extent that structured flowcharts are increasingly used as the basis for Computer-Aided Software Engineering (CASE).

Seven types of flowcharts will be discussed below. The Standard Process Flowchart is the most widely used, probably because it has the broadest applicability to problem solving and because it can be used as a baseline to create some of the other flowcharts that may better graphically represent a particular point. You should modify your use of flowcharts to your specific needs. Be creative.

An important part of flowcharts are the symbols used to represent various kinds of activities. The American National Standards Institute (ANSI) has developed a standard set of symbols for flowcharting. We have seen many modifications of the ANSI symbol set used effectively in organizations. The important thing is that within an organization there is consistency among documents and, therefore, consistency in the "symbology" used. The Forms part of this section displays a list of some of the common ANSI symbols that you can use as a starting point.

Steps

There is really no prescribed sequence for generating flowcharts. However, some commonsense rules should apply in flowcharting. The following general steps have proven to be effective in most flowcharting activities:

1. Before beginning, define your objectives for flowcharting.
2. Determine the process boundaries:
    a. What is included in the process?
    b. What is not included?
    c. What are the outputs from the process?
    d. What are the inputs to the process?
    e. What departments are involved in the process?
3. Select the most appropriate flowchart type(s) for your objectives.
4. Prepare a high-level block diagram of the process you wish to flowchart. Stretch the boundaries of the process at this point to get the broadest view of the process possible. Prepare a block diagram regardless of if you will ultimately use block diagrams or different flowchart types.
5. Determine what people or functional areas are involved in the process. Assemble the most appropriate team to flowchart the process.
6. Determine what tools and/or standard formats will be used, if this has not already been established by a broader, perhaps companywide, initiative.
7. Begin flowcharting the process. Start at a high level and work toward more detail. Pay close attention to all interactions among people, departments, and functions. Using the guidelines and tips below, identify the suppliers of all process inputs and the customers of all process outputs.
8. Early in the creation of flowcharts, review the appropriateness of the process boundaries. Modify the process definition, based on its boundaries, as necessary. Also, modify the team members, if necessary, to reflect any changes in process definition.
9. Complete the flowcharts.

Guidelines and Tips

The level of detail of your flowcharts is largely a matter of judgment and this will improve with experience. By fully understanding your objectives before starting to flowchart, you will have a better idea of the meaningful level of detail. Early in the flowcharting activity, it is often a good idea to early on take a small part of the process and dissect it with the team. Take it to a very fine level of detail and then agree with the team on the appropriate level of detail, based on the team's objectives.

There are two fundamentally different ways to approach flowcharting. One way is to start at the beginning and define each step in great detail. Another way is to build a simple view of the process first and then add detail. While you should ultimately choose the method you are most comfortable with, our experience has been that it is generally easier to start simple and add detail.

It is very important to separate the flowcharting process from process improvements. Focus first on completing flowcharting. Create an "Issues List" and defer judgment on this list until flowcharting is complete. Discussions on potential improvements can be a great distraction and delay the overall improvement process.

It is very important to flowchart processes as they are today or to flowchart a vision of the future. It is crucial not to mix these in the same flowchart. Effective flowcharting, therefore, may require interviewing skills to uncover the actual processes used today. Flowcharting may also be supplemented with process walk-throughs to ensure that current processes are accurately captured. Walk-throughs are a highly recommended part of the process. You will be surprised how many things you find during a walk-through that are different from the way you have documented the process. The steps to a process walk-through are as follows:

1. Identify the scope of the process to be reviewed.
2. Develop the objectives of the walk-through.
3. Determine the walk-through method:
    a. Interview
    b. Observation
    c. Sampling
4. Create the interview worksheet (who, what, when).
5. Conduct the walk-through.
6. Update the flowchart.

Some of the discoveries of process walk-throughs include:

- Differences between the documented process and present practice
- Differences among employees in the way they perform the activity
- Employees in need of retraining
- Suggested improvements to the process, identified by the people performing the process
- Activities that need to be documented
- Process problems, such as:
  - Duplication
  - Rework
  - Waste
  - Bureaucracy
- Roadblocks to process improvement
- New training programs required to support the present process

Put dates on all flowcharts created. Flowcharting is an iterative process. Including dates on all revisions will save you many headaches later on.

Example

See Figures 19.1 to 19.5.

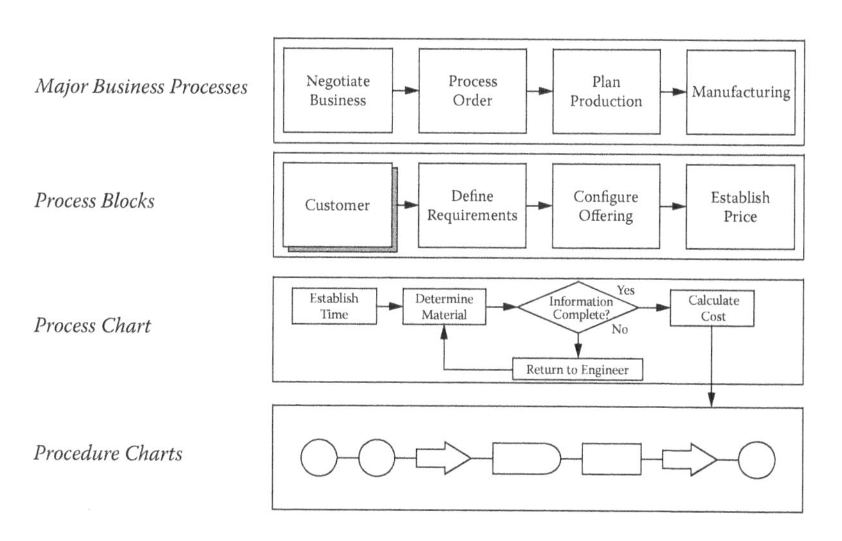

FIGURE 19.1
Relationships among flowcharting techniques.

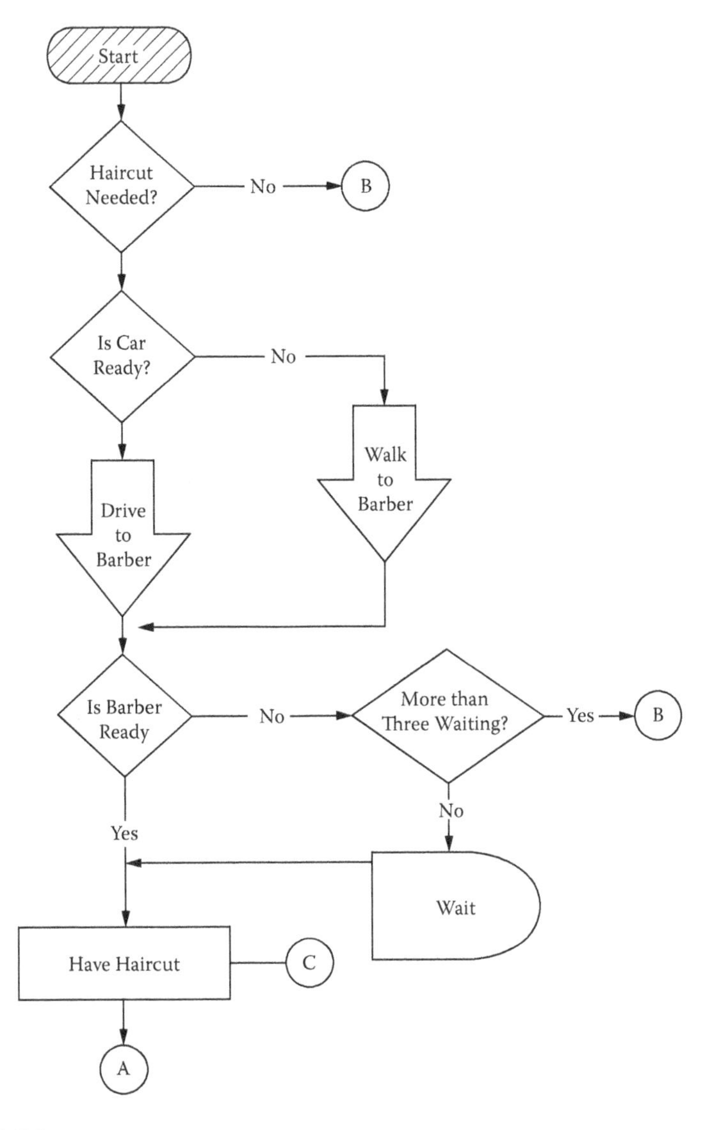

FIGURE 19.2
Two standard process flowcharts of different parts of the process of getting a haircut and/or going fishing (Part I).

Software

There are currently a number of software products on the market that can assist you in creating process flowcharts. Some of them are generic tools, such as drawing programs, while others have been created specifically for flowcharting. One product we suggest is a software package produced by

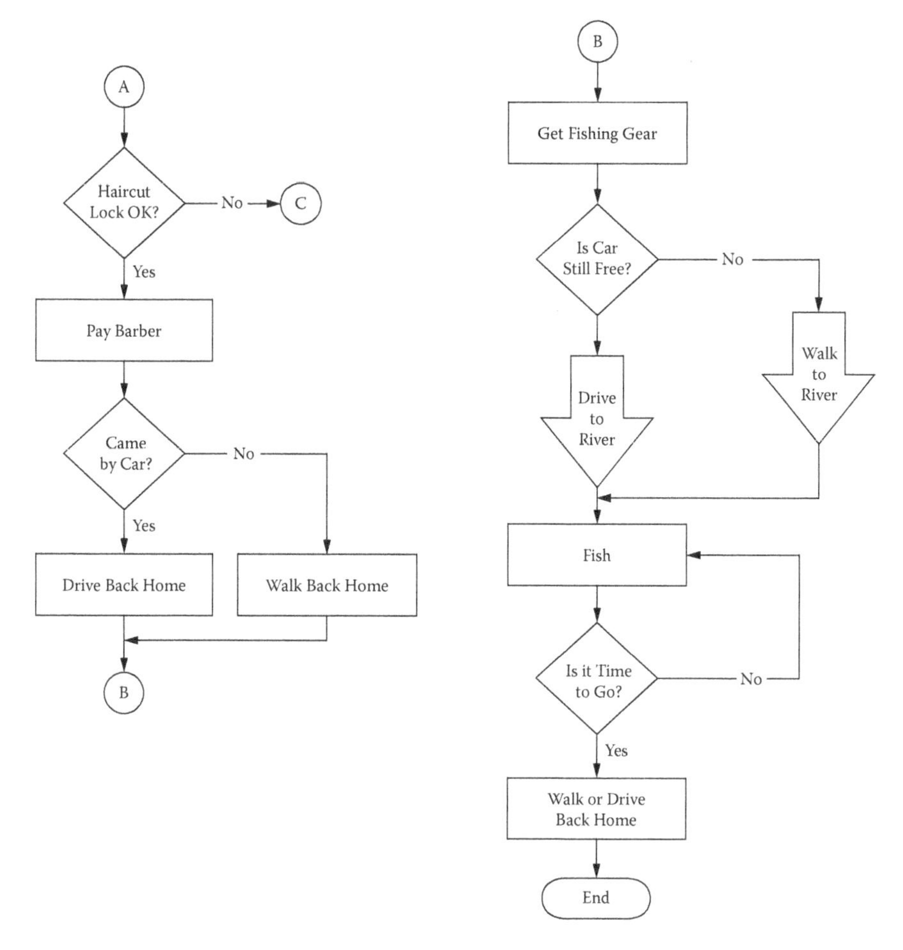

FIGURE 19.3
Two standard process flowcharts of different parts of the process of getting a haircut and/or going fishing (Part II).

Edge Software Inc. (Franklin, WI), called WorkDraw. This is an advanced process modeling application that includes flowcharting capabilities.

- Visio 14.0.6
- SmartDraw® VP 19.1.3.2
- FlowCharter® 14.1.2
- Edraw Flowchart 6
- EDGE Diagrammer 6.24
- WizFlow Flowcharter 6.24
- RFFlow 5.06

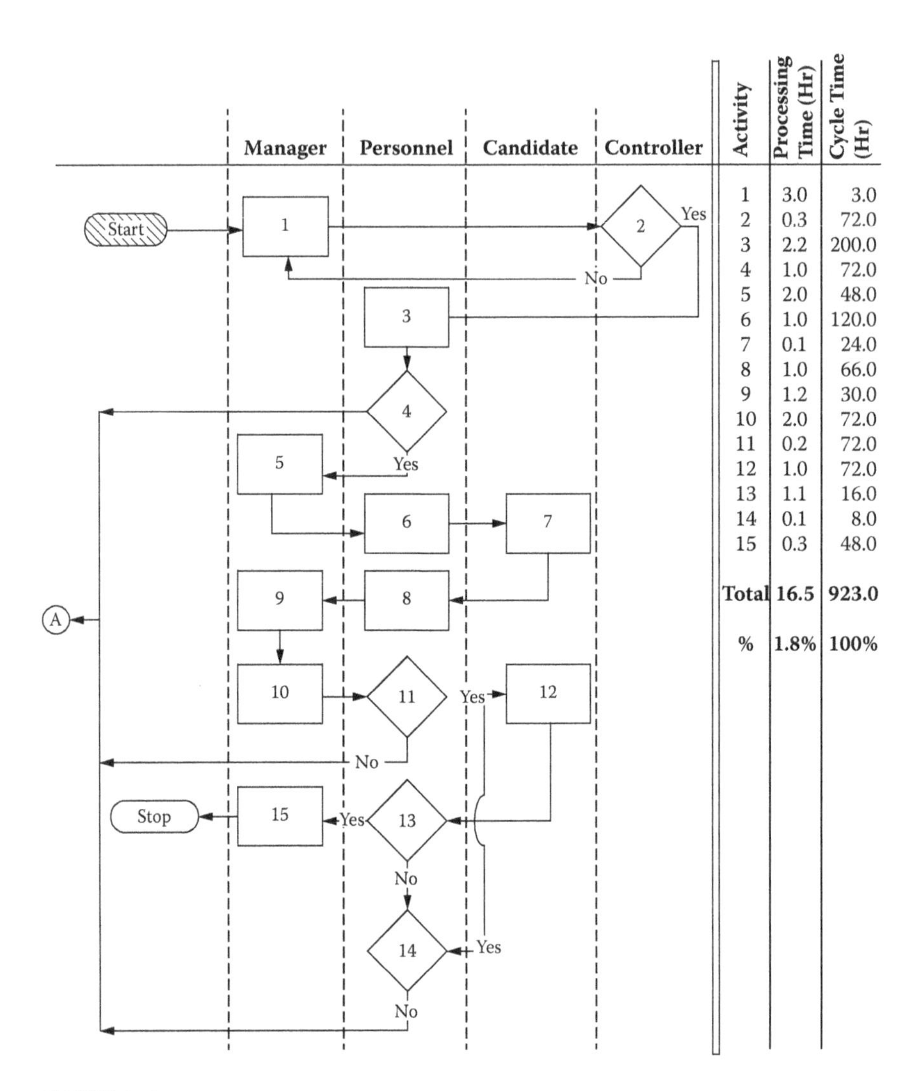

Activity	Processing Time (Hr)	Cycle Time (Hr)
1	3.0	3.0
2	0.3	72.0
3	2.2	200.0
4	1.0	72.0
5	2.0	48.0
6	1.0	120.0
7	0.1	24.0
8	1.0	66.0
9	1.2	30.0
10	2.0	72.0
11	0.2	72.0
12	1.0	72.0
13	1.1	16.0
14	0.1	8.0
15	0.3	48.0
Total	**16.5**	**923.0**
%	**1.8%**	**100%**

FIGURE 19.4
Functional flowchart of the internal job search process.

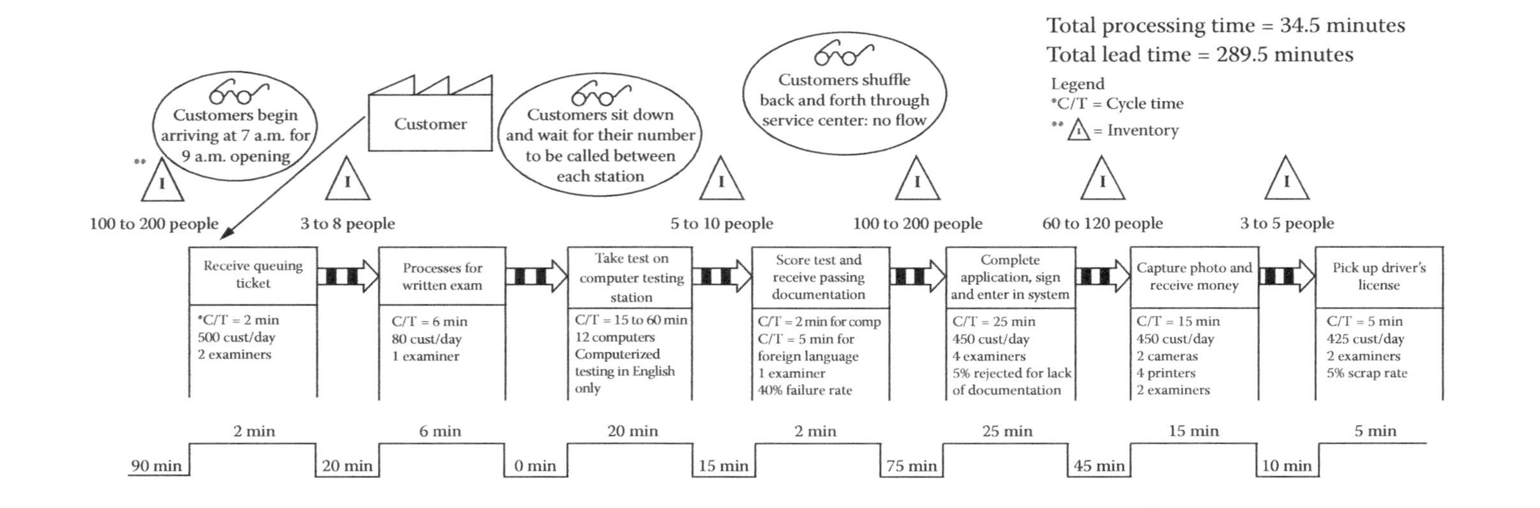

FIGURE 19.5
Driver's license current state map. (From *Quality Digest Magazine*, March 2006, page 41. With permission.)

ADDITIONAL READING

Esseling, E. K. C., H. J. Harrington, and H. VanNimwegen. 1997. *Business process improvement workbook*. New York: McGraw-Hill.

Galloway, D. 1994. *Mapping work processes*. Milwaukee, WI: ASQ Quality Press.

Harrington, H. J. 1997. *The business process improvement workbook*. New York: McGraw-Hill.

Harrington, H. J., G. Hoffherr, and R. Reid. 1998. *Area activity analysis*. New York: McGraw-Hill.

PQ Systems, Inc. 1996. *Total quality tools*. Milwaukee, WI: ASQ Quality Press.

20

Force Field Analysis

DEFINITION

Force field analysis is a method to help identify the positive and negative forces working on a process when trying to attain a new state. It is a visual aid for pinpointing and analyzing elements that resist change (restraining forces) or push for change (driving forces). This technique helps drive improvement by developing plans to overcome the restrainers and make maximum use of the driving forces.

Its Uses

Force field analysis results in a well-defined and well-understood goal. This goal can be used to stimulate a brainstorming of solutions for the improvement process. It also can help to determine activity root causes.

General

Force field analysis is used to help better understand a situation. The premise is that the current state of a situation is a balance between forces driving the situation toward a desired state and forces restraining movement toward that desired state. Examination of these two types of forces allows for more effective management of change. The amount and strength of each type of force gives one a basic idea of the difficulty in reaching the desired state. The technique not only helps to identify driving and restraining forces, but it also assists in generating solutions.

Preparation

Applying force field analysis:

1. Define the goal or desired state
2. Define the current situation
3. Identify the restraining forces
4. Identify the driving forces
5. Determine which forces, if any, are substantially stronger than the rest
6. Determine the feasibility of changing any of the forces
7. Develop and implement a strategy for changing those forces
8. Reexamine the situation

Example

- Define the goal or desired state.
  Write down the goal on top of a white board or page.
- Define the current situation.
  Draw a line across the middle of the board. Label this line with the current situation (Figure 20.1).
- Identify the restraining forces.
  Identify the restraining forces that exist and are keeping the desired state from being achieved. Label these forces as arrows pushing down on the current state line (Figure 20.2).
- Identify the driving forces.
  Identify the driving forces that help to move the current state toward the desired state. Label these forces as arrows pushing up on the current state line (Figure 20.3).
- Determine which forces, if any, are substantially stronger than the rest.
  Make the arrows representing these forces larger (Figure 20.4).

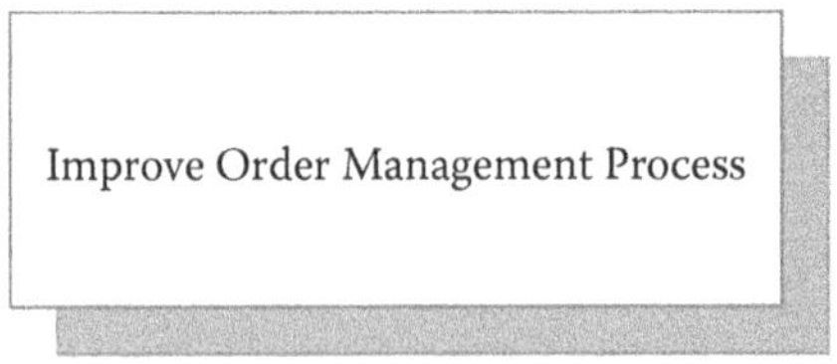

Improve Order Management Process

FIGURE 20.1
Your goal: Current situation.

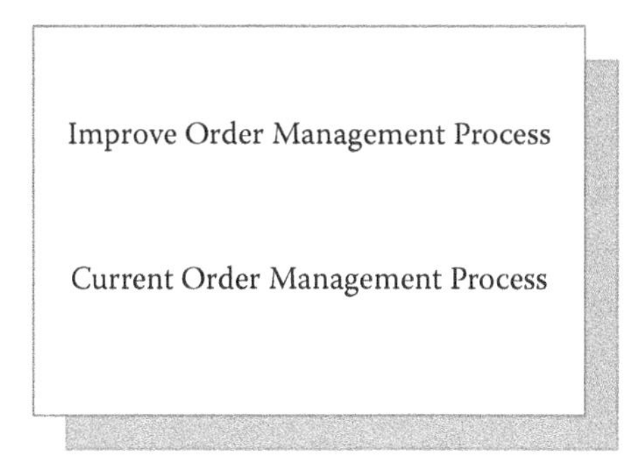

FIGURE 20.2
Your goal: Current situation–3.

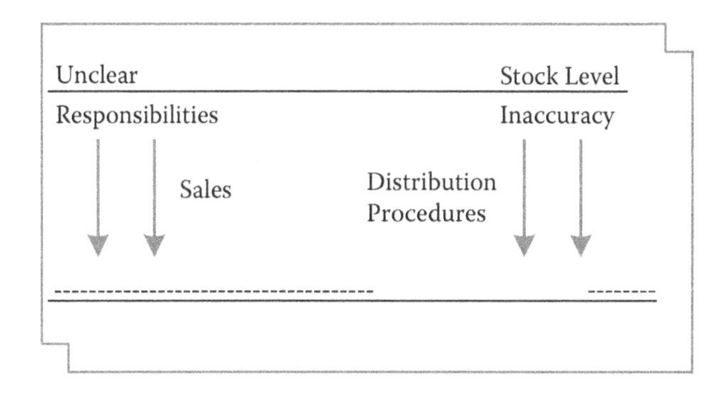

FIGURE 20.3
Your goal: Current situation–4.

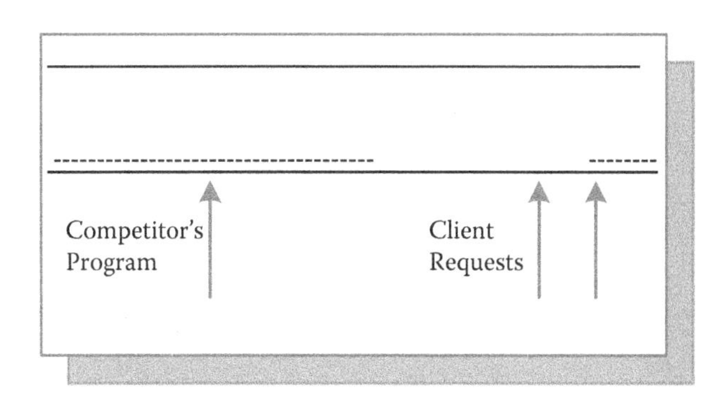

FIGURE 20.4
Your goal: Current situation–5.

- Determine the feasibility of changing any of the forces.

  Determine the feasibility of changing any of the forces, giving the strongest forces more priority. Feasibility is determined by ease of change and potential impact.
- Develop and implement a strategy for changing those forces.

  Develop and implement a strategy for changing those forces that need to be and can be changed. This consists of three alternatives: strengthen driving forces, reduce or eliminate restraining forces, or create new driving forces. Elimination of restraining forces is typically the most effective method.
- Reexamine the situation.

  Reexamine the situation to determine the effectiveness of any implemented changes. Make adjustments as necessary (Figure 20.5 and Figure 20.6).

Software

Some commercial software available includes, but is not limited to:

- SmartDraw®

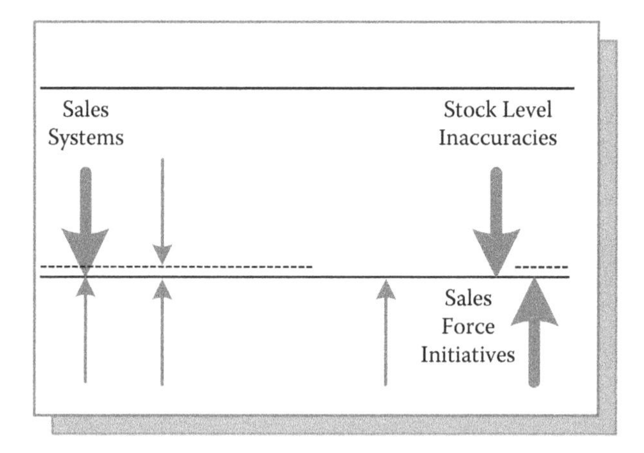

FIGURE 20.5
Force field analysis with both positive and negative forces indicated.

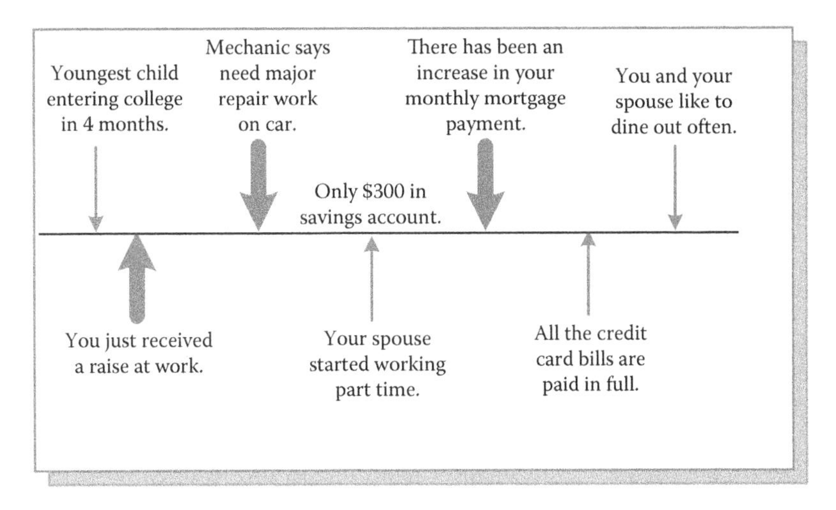

FIGURE 20.6
Typical force field graphic presentation.

21

Graph

DEFINITION

Graphs is a method for visually comparing two or more sets of data. Graphs are visual displays of quantitative or qualitative data. They visually summarize a set of numbers or statistics.

Its Uses

Graphs are used for interpretation of many facts, especially statistical data, which are often best attained by analyzing graphs.

General

Graphic presentation does the most to simplify interpretation. Most graphs begin with a construction of the *X* and *Y* axes.

Preparation

X–Y Axes Graphs

A graph is a pictorial representation of data on sets of horizontal and vertical lines called a grid. The data are plotted on the horizontal and vertical lines. Each axis is assigned specific numerical values corresponding to the data. The horizontal line is called the *X* axis while the vertical line is the *Y* axis. The point where the two lines meet is called the *zero point* or *point of origin*.

The time frame or cause of data is plotted on the *X* axis (horizontal); they also are referred to as the *independent variable*. The effect of variation in the independent variable over a period of time is plotted on the

Y axis (vertical). This is referred to as the dependent variable, as the values vary in relationship to the independent variable. If two variables depend on each other, or if both are affected by some other factor, they can be arbitrarily placed on either axis (Figure 21.1).

Line Graphs

Line graphs are the simplest graphs to prepare and use. They show the relationship of one measurement to another relative to the independent variable or over a period of time. Often this graph is continually created as measurement occurs. This procedure may allow the graph to serve as a basis for projecting future relationships of the variables being measured (Figure 21.2).

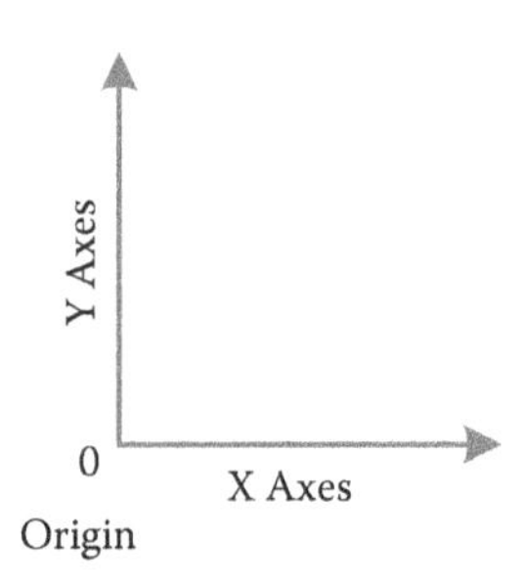

FIGURE 21.1
X–Y graph.

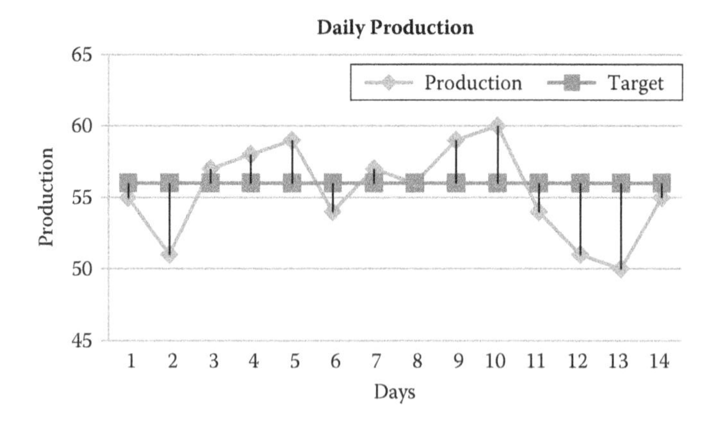

FIGURE 21.2
Line graph.

Multiple Line Graphs

When there are more than two sets of data related to the same independent variable, they can be plotted on a multiple line graph (Figure 21.3).

Bar or Column Graphs

Bar graphs have horizontal (bars) or vertical (columns) that show their relationship to each other relative to an independent variable. The bars and columns may show two or more related measurements in several situations, or different categories, which show the cumulative measure. Bar graphs have the bars originating from the *Y* axis. As a consequence, the normal location of dependent and independent variables is reversed (Figure 21.4).

Area Graphs

Area graphs show how 100 percent of something is assigned. The most commonly used area graph is the pie chart (Figure 21.5).

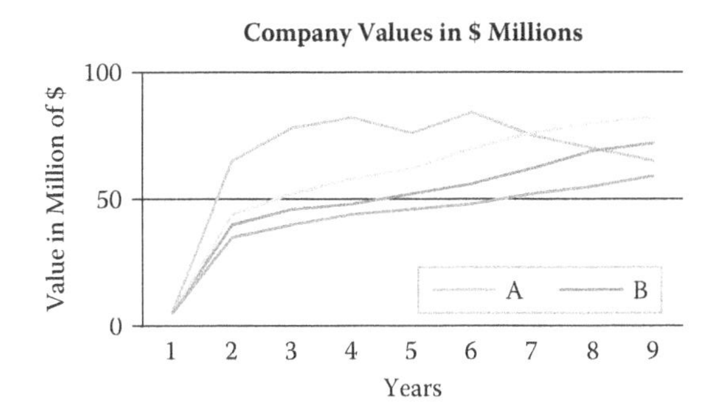

FIGURE 21.3
Line graph.

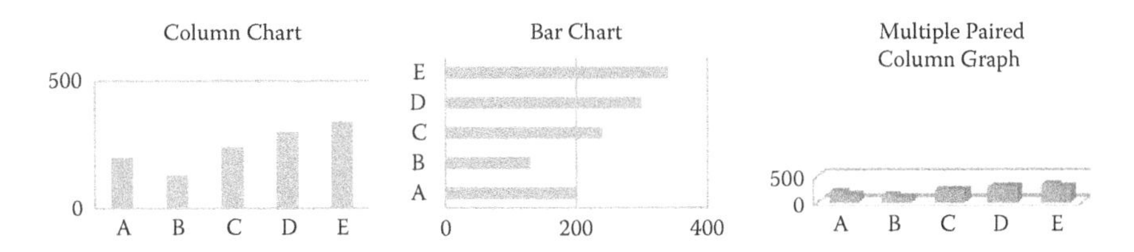

FIGURE 21.4
Bar graphs.

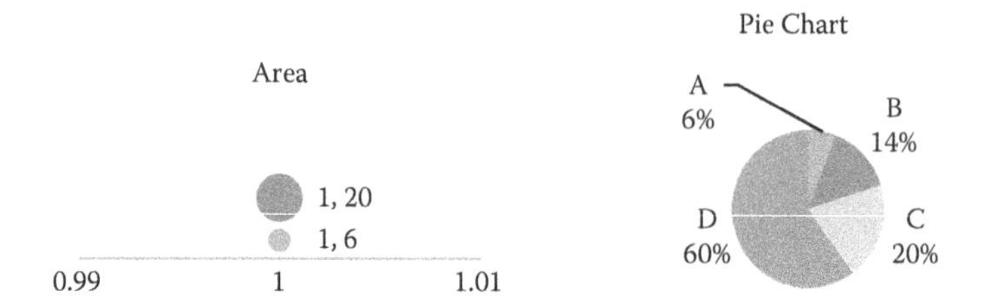

FIGURE 21.5
Area graphs.

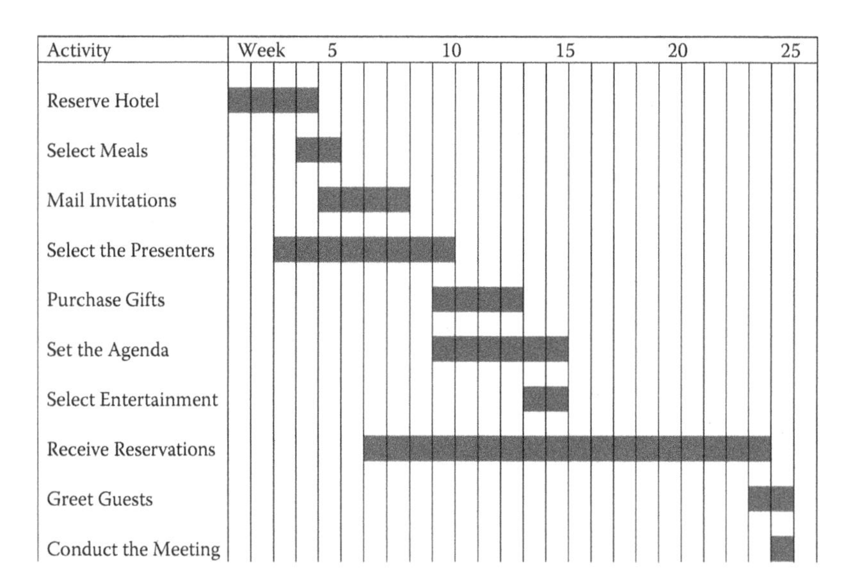

FIGURE 21.6
Gantt charts.

Gantt Charts (also referred to as Milestone or Planning Graphs)

These charts show the goals or target to be achieved by depicting the projected schedule of the project or process. A primary purpose is to help organize projects and to coordinate activities (Figure 21.6).

Pictorial Graphs

Pictorial graphs use pictures or drawings to represent data. A pictogram is a type of pictorial graph in which a symbol is used to represent a specific quantity of the item being plotted. The pictogram is constructed and used like area, bar, and column graphs (Figure 21.7).

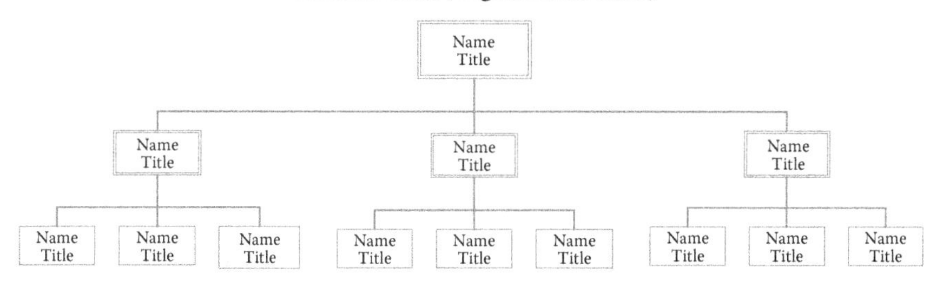

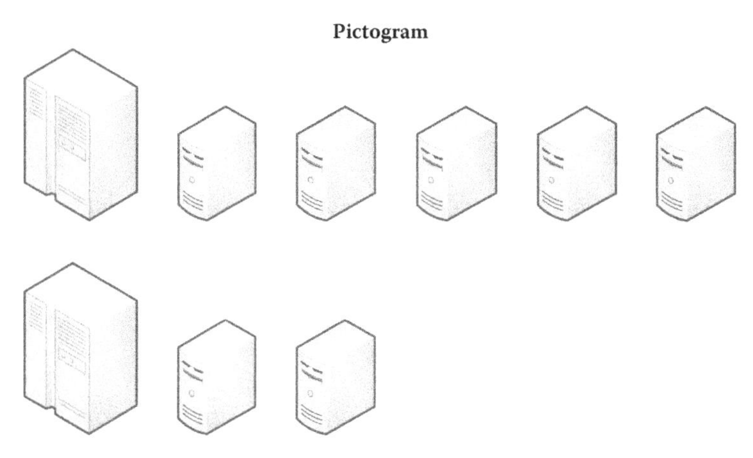

FIGURE 21.7
Pictorial graph.

Example

One was already presented in the preparation part of this tool.

Software

Some commercial software available includes, but is not limited to:

- Excel®
- Lotus 1-2-3®
- Minitab®
- JMP®
- RFFlow

22

Kano Model

DEFINITION

The Kano model is a theory of product development and customer satisfaction, which classifies customer preferences into five categories:

- Must-be quality
- One-dimensional quality
- Attractive quality
- Indifferent quality
- Reverse quality

Its Uses

The Kano model is used to gain an understanding of a customer's needs. It is a useful tool in the following activities:

- New product development
- New service development
- Continued product improvement
- Determining market strategies

General

The model is plotted in two dimensions:

- Degree of implementation along the horizontal axis:
  - Low is "not done"
  - High is "fully implemented"

- Degree of satisfaction along the vertical axis:
  - Low is "dissatisfied"
  - High is "satisfied"

The elements of the plot include:

- Attractive: Customers are satisfied when achieved fully, but do not cause dissatisfaction when not fulfilled. These are attributes that are not normally expected. For example, a car door includes a light to illuminate the ground when opening it at night. These types of delighters are not often identified by customers.
- One-dimensional quality: Customers are satisfied when these are fully met and dissatisfied when they are not met. These are attributes that are identified by customers and which companies include in their products in order to compete with competition. An example would be a light in the car visor that illuminates when the mirror is exposed. Not needed, but when wanted, it must be there.
- Must-be quality: Customers expect these to be present and when they are not present, they are dissatisfied. An example would be a light in a refrigerator that lights when the door is opened. Customers consider these as basic and will not tell the company if it is not there, they simply will not buy.
- Indifferent quality: Customers are neither satisfied nor dissatisfied if these are present or missing.
- Reverse quality: All customers are not alike. Some customers do not want advanced technology; rather they prefer the basic product. For example, smartphones versus simple cell phones. Some customers prefer high-tech while others prefer the basic functions and will be dissatisfied with the high-tech features.

Preparation

- Data are collected by several means, for example:
  - Customer interviews
  - Market evaluations
  - Expectations
  - Future items

- The product or service is rated against these items and the performance of the product or service is graphed.

Example

See Figure 22.1.

Software

Some commercial software available includes, but is not limited to:

- Pragmatic Marketing®

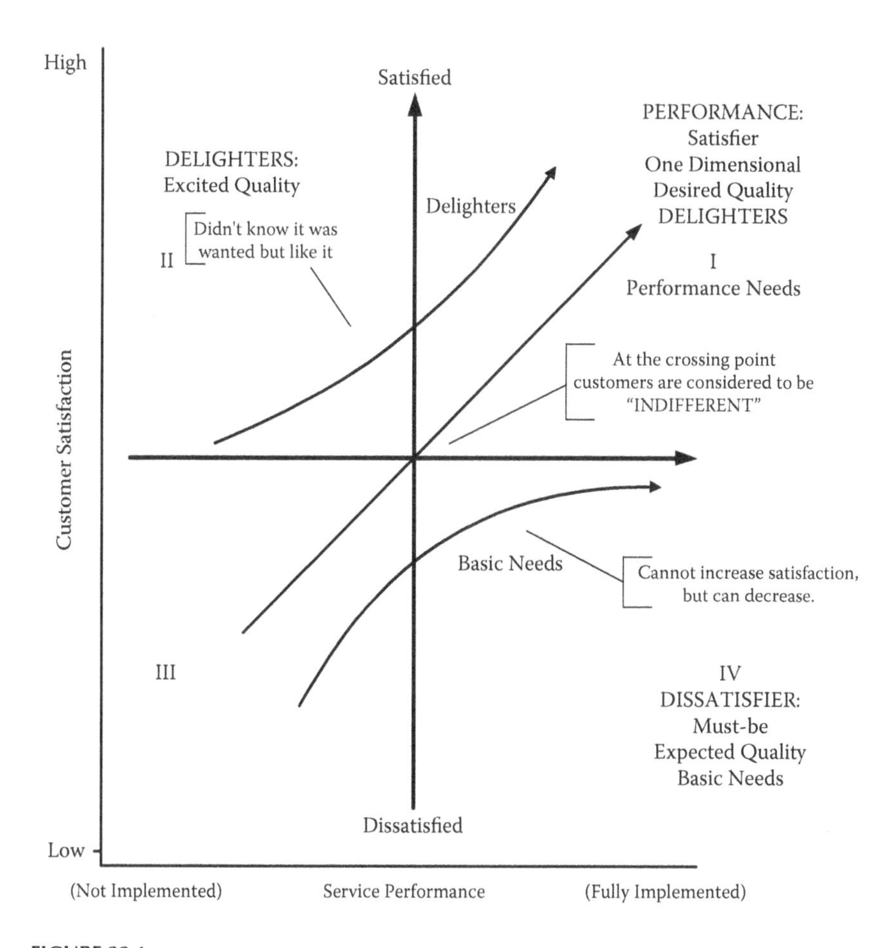

FIGURE 22.1
Kano model.

23

Motivation Management

DEFINITION

Motivation management is developing an understanding of what provides an individual with a sense of self-worth, pride, and accomplishment when using this understanding to guide the way work is assigned and how individuals are recognized for their efforts. It results in an organization developing its values, beliefs, procedures, and culture in a way that drives the organization's employees to get more enjoyment out of the work they are doing and be more committed to the organization. It makes use of a set of logical neuro programs that suit the perception of a person's or an organization's needs for the purpose of efficiency and optimality to accomplish desired organizational goal.

Its Uses

Motivation is used to inspire people and groups to produce the best results in the most efficient and effective manner. Motivation is unique to each individual; it can't be dictated from the outside. Therefore, it is important that management learns how to identify and address these motivating forces.

Motivation management's goal is to increase production and efficiency to reach maximum results for the organization. Motivation for better performance depends on:

- job satisfaction
- achievement
- recognition
- professional growth (Boyett and Boyett, 2000)

It is only by understanding the professional needs of an individual that a positive motivational work environment can be achieved.

General

Organizational change, by definition, creates resistance. This is because disruption is usually what people are resisting more than the change itself. Disruption is the focal point in understanding resistance. The fear of ambiguity and loss of control during change is powerful enough to immobilize many people and prevent their movement to even highly desired circumstances. Substantive change can occur only when people feel that they have no choice; the energy of resisting the change becomes greater than the energy of doing the change.

Because of these psychological dynamics, managers must understand, appreciate, and apply the basic concepts and techniques of pain management. Pain management is the process of consciously surfacing and orchestrating certain information in order to generate the appropriate level of discomfort regarding the continuance of the status quo.

Preparation

Change-related motivation or pain refers to the level of discomfort a person experiences when his or her goals or expectations are not being met (current pain) or are not expected to be met (anticipated pain) due to the status quo. This pain occurs when people pay or will pay the price for an unresolved problem or miss a key opportunity (Figure 23.1).

Key concepts of motivation management include:

- People are generally very frightened of the transition state and will avoid major change if at all possible.
- The only time major change occurs is when the cost or pain of the status quo is greater than the cost of transition to the future state.
- A critical mass of this type of pain must be present to justify a person breaking the inertia of their status quo.
- The pain threshold for change occurs when the cost of the status quo is greater than the cost of transition to the future state.
- The level of pain needed to achieve this critical mass is relative to two factors:
  - Frame of reference
  - Pain tolerance

	Problem	Opportunity
Current Situation	SITUATION: A problem exists. PAIN: Loss of: • Job security • Organizational survival • Market share • Etc.	SITUATION: Immediate action can mitigate or eliminate the problem. PAIN: Loss of potential advantage
Anticipated Situation	SITUATION: We will be in trouble. PAIN: Impending loss of: • Job security • Organizational survival • Market share • Etc.	SITUATION: The future state will be better than where we are today. PAIN: Loss of future advantage for achievement

FIGURE 23.1
Situation analysis matrix.

Pain does not generate change when it is:

- Denied: Denial occurs when the evidence of pain is obvious to others, or that the person in question blocks out the information from conscious awareness. Generally, such a person cannot afford to consciously acknowledge the pain because the problem or solution or both is too threatening. "We don't have a major problem. The decline in productivity is only temporary."
- Ignored: If the pain is acknowledged, but not perceived as significant enough to merit any real attention, the response is either: "It's no big deal" or "It's a problem, but it will work itself out."
- Delayed: Change-related pain is a conscious act while experience also involves the subconscious mind. "I see it, but I cannot respond." Either the person doesn't know what to do, or doesn't want to pay the price it will take for resolution. He/she feels there is no feasible alternative, so he/she feels trapped.

Change-related pain can be generated from perceptions about the past, present, future, ambiguity, cost, and levels of risk associated with the status quo. By surfacing dissatisfaction with the status quo, change-related pain strategies can be tailored to the client. Some tactics to address pain include:

- Cost/benefit analysis
- Business imperative analysis
- Industry benchmarking
- Cost of quality audit
- Customer survey
- International quality study
- Industry trend analysis
- Audit management letter
- Readiness assessment
- Force field analysis

Steps in motivational management include:

1. Arrange a meeting with the key executive sponsors.
   Arrange a meeting with the key executive sponsors to measure the sponsors' awareness of the need to move from the status quo.
2. Determine the frame of reference.
3. Identify hot spots and levels of comfort. Utilizing the Organization Change Management (OCM) tool, motivation management strategies can be developed to identify hot spots and levels of comfort with the project.
4. Formulate a presentation to the Executive Sponsor Team.
   Formulate a presentation to the Executive Sponsor Team concerning the need to move from the status quo and the realistic disruption that this transition will create.
5. Repeat steps.
   Repeat steps 1 through 4 with representative targets utilizing the OCM tool: Motivation management strategies: Target.

Example

Typical management action to show employees that the organization appreciates a job well done:

- Employee of the month recognition
- Presentations to upper management by the employee

- Dinner for two to recognize unusual extra effort
- Pizza and drinks at a department meeting to recognize a major accomplishment that the total department was involved in
- Town meetings to share progress, accomplishments, and challenges

Software

Some commercial software available includes, but is not limited to:

- Halogen®

REFERENCE

Boyett, J. H., and J. T. Boyett. 2000. *The guru guide: The best ideas of the top management thinkers*. Hoboken, NJ: John Wiley & Sons.

24

Negative Analysis

DEFINITION

Negative analysis is an approach used to look at a process or situation to define what action could be taken to cause a negative impact upon the results. It generates a list of actions that, if implemented, would result in making the present situation worse. It then generates action plans to minimize the impact that these actions would have upon the process or situation.

Its Uses

Negative analysis is a method used to define potential problems before they occur and developing countermeasures.

General

This technique defines risks related to a process or situation and develops action plans that will minimize the impact these risks would have upon the process or situation. It is often used in conjunction with the development of a future state solution in order to identify potential problems related to the projected changes before they are implemented.

Preparation

- Identify and quantify the process or situation to which negative analysis will be applied.
- Organize a group of knowledgeable people to participate in the negative analysis session.
- Develop the ground rules for the session.

- Tell the group which method of participation will be used. Choose either a structured or unstructured approach:
  - A structured approach allows everyone in the group to give their ideas as their turn arises in the rotation or pass until the next round. The structured method allows every member of the group an equal chance to participate.
  - An unstructured approach allows the group to give ideas as they come to mind. The unstructured method creates a more relaxed atmosphere, but also risks unequal participation from the group.
- Frame the challenge:
  - Take two minutes as a group to think silently.
  - Call out and record ideas; don't allow any discussion or judgment.
  - Review the lists classifying each one's probability of occurrence as high, medium, or low.
  - For all the ones rated as high potential of occurrence, develop and implement action plans to offset these situations. For those with a medium level potential of occurrence, generate action plans and determine which of the plans are practical for implementation. Unless it is an extremely critical or safety-related low probability of occurrence level, it is usually best to realize the situation could occur, but it is usually not necessary to generate action plans to prevent the situation from occurring at this point in the project.

Example

Let's assume that we have invested in a steak restaurant, but, strangely enough, we wanted it to fail. What would we do to make sure that it will go bankrupt? The following are some typical actions that could put our restaurant out of business:

- Don't put up a sign advertising the restaurant.
- Have the employees park in all of the parking spots.
- Keep people waiting a minimum of 15 minutes even if there are tables available.
- Be sure that there is lipstick on every glass when it is placed on the tables.
- Open the restaurant from 10 a.m. to 2 p.m. only.
- Serve steaks from milk cows only (very tough and lean).
- Don't serve beer, wine, or hard liquor.

This list could go on and on. However, let's see what we could do to offset some of these negative impact conditions:

- To offset condition 1, we can put a very large, brightly lit sign right in front of the restaurant.
- To offset condition 2, we can restrict our employees from parking immediately adjacent to the restaurant and instead require them to park in a lot across the street.
- To offset condition 3, we can have several hostesses on staff so that, as people enter the building, they are ushered immediately to a table. If there is no table available, then the hostess should provide a legitimate, but very conservative estimate (always estimate longer than it should take, not shorter) of when they will be seated and check back with them every five minutes to tell them that you are still trying to seat them as soon as possible.
- To offset condition 4, make sure that the dishwashers can remove lipstick from all glasses no matter how thick the lipstick is or what kind of lipstick it is.

This provides you with some examples of how negative analysis could work. It could be a lot of fun. Enjoy it. In the long run, you will be surprised how effective it is.

ADDITIONAL READING

Harrington, H. J. 1991. *Business process improvement.* New York: McGraw-Hill.

Harrington, H. J. 2013. *Streamlined process improvement.* New York: McGraw-Hill.

Harrington, H. J., and J. S. Harrington. 1995. *Total improvement management.* New York: McGraw-Hill.

25

Nominal Group Technique

DEFINITION

Nominal group technique is a special purpose technique, useful for situations where individual judgments must be tapped and combined to arrive at decisions. It is a process to develop and narrow alternatives by generating ideas.

Its Uses

Nominal group is used to develop and narrow alternatives by generating ideas and using a voting process.

General

By allowing for equal involvement of all group members, this technique creates high-quality alternatives, effective consensus building, and a strong feeling of group accomplishment.

Preparation

Using the Nominal Group Technique

1. Generate lists of responses.

   Instruct participants to write on a piece of paper their responses to the problem or task statement. Do not allow participants to discuss their responses.
2. Organize the participants:
   a. Pull together a group of 7 to 10 individuals.
   b. Select a leader or facilitator from among the members of the group.

3. Introduce the session.

   Describe the purpose of the session. Typically this will include a thorough and clear presentation of the problem or task statement. It is also a good idea to describe the nominal group process to the participants.
4. Use a round robin process to present ideas to the entire group.

   Ask each participant to take turns stating one written response each time around, until all responses have been presented. Limit discussion to the participant who proposed the idea and the person recording the ideas.

   Participants may pass at any time and contribute next time around. Encourage participants to add ideas to their personal lists as new ones arise.
5. Clarify the ideas.

   Briefly discuss each idea with the group for clarification purposes only. New ideas may be generated at this time by combining or modifying previously generated ideas. Do not permit the evaluation of any idea, new or old. Eliminate duplicate ideas.
6. Rank the ideas.

   Ask each participant to rank all ideas in his or her personal order of importance or preference.

   Ranks should be assigned according to the number of ideas generated, and the highest ranked idea should receive the highest number. For example, if 13 ideas are generated, the participant should assign a "13" to his or her favorite idea and a "1" to the least favorite idea.
7. Tally the ranking votes.

   Total the ranking scores for each idea. The idea with the highest point total is the one that the entire group has given highest priority.

   Look for peculiar ranking patterns, such as when an idea receives many high and low priority rankings, but none in the middle, or the majority of rankings fall in the middle, and so on. Tallies such as these may indicate a problem in clarifying the idea or simply a different interpretation across participants. Any such anomaly may require a reinterpretation of the ideas, followed by another ranking session.

Example

The team is presented with the following task statement: "Our administrative group cannot keep pace with the required documentation generation." Each team member would document his or her response on the Ideas Worksheet. Team member "Bob" shows the following information on his worksheet (Figure 25.1).

The full list of Bob's ideas and those of his team are recorded on a flipchart (Table 25.1).

After our team's idea list has been clarified and narrowed down, it is time for Bob and the rest of the team to rank the remaining ideas. Bob's ranking form is presented in Figure 25.2.

A new team list of "preferred" ideas is generated. Very often this will consist of each team member's "first and second" choices (Figure 25.3).

Nominal Group Technique Worksheet: Ideas	
Problem Statement:	**Date:** January 5, 1998
"Our administrative group cannot keep pace with the required documentation generation."	
Individual Ideas:	
Hire more people	
Distribute the work more evenly	
Take in fewer new tasks	

FIGURE 25.1
Bob's ideas worksheet.

TABLE 25.1
Team's Idea List

- Hire more people
- Distribute the work more evenly
- Take in fewer new tasks
- Change from word processors to computers
- Hire part-time workers
- Use management to fill in
- Take fewer breaks
- Take more breaks
- Buy new word processors
- Provide better training
- Provide any training
- Provide comfortable chairs
- Designate workflow supervisors
- Use first-in–first-out system
- Outsource some of the work

Nominal Group Technique Worksheet: Ranking	
Problem Statement: "Our administrative group cannot keep pace with the required documentation generation."	**Date:** January 5, 1998
Idea	**Ranking**
A. Hire more people	7
B. Distribute the work more evenly	8
C. Take in fewer new tasks	9
D. Outsource some of the work	6
E. Change from word processors to computers	1
F. Hire part-time workers	3
G. Provide better training	4
H. Use first in–first out system	5
I. Buy new word processors	2

FIGURE 25.2
Bob's ranking worksheet.

Team's Preferred Ideas
• Change from word processors to computers
• Buy new word processors

FIGURE 25.3
Team's preferred ideas.

Once again, it is time for Bob's team to reach consensus on which of the solutions to go forward with. In this case, the team selected to change from word processors to computers as a solution to their issue.

As you can see, nominal group technique uses several other basic approaches, such as brainstorming, narrowing techniques, and consensus. This approach is very simple to use, yet is dynamic in its results.

Software

Some commercial software available includes, but is not limited to:

- WinSite!®

ADDITIONAL READING

Brassard, M. 1989. *The memory jogger plus.* Milwaukee, WI: ASQ Quality Press.

Harrington, H. J. 1987. *The improvement process: How America's leading companies improve quality.* New York: McGraw-Hill.

Lynch, R. F., and T. J. Werner. 1991. *Continuous improvement team and tools: A guide for action*. Milwaukee, WI: ASQ Quality Press.
Rieker Management Systems. 1986. *Systematic participative management—Team member manual.* San Jose, CA.
Tague, N. R. 1995. *The quality toolbox*. Milwaukee, WI: ASQ Quality Press.

26

Organizational Change Management

DEFINITION

Organizational change management (OCM) is a methodology designed to help prepare the organization and the individuals within the organization to accept changes to the organizational structure, processes, and operating procedures. It is designed to break down resistance to change and to build up organization resiliency.

Its Uses

Organizational change management is used when organizational change is required in order to address business situations that the current organizational structure cannot handle.

General

One of the things that distinguishes a successful person from an unsuccessful person is his/her ability to deal with change. Successful people develop the capacity to really understand and evaluate the alternatives available to them. They take advantage of the change activities and get involved in them so that they can influence the outcome.

Preparation

- Ask direct questions about who the sponsor is and who the less visible parties to the contract are.
- Elicit the sponsor's expectations of you.

- Clearly and simply state what you want from the sponsor.
- Say no or postpone a project that in your judgment has less than a 50/50 chance of success.
- Probe directly for the sponsor's underlying concerns about losing control.
- Probe directly for the sponsor's underlying concerns about exposure and vulnerability.
- Give direct verbal support to the sponsor.
- When the contracting meeting is not going well, discuss directly with the sponsor why this contracting meeting is not going well.

Within the context of OCM contracting, all contracts are written documents. An OCM contract can simply be an agreement to perform some act, meet a deadline, or behave in a certain manner. The primary reason to put a contract in writing is to clarify key points in the relationship and to communicate your mutual understanding of the agreement. Writing down the agreement forces more explicit detail concerning what you are going to do. A contract should be brief, direct, and conversational. It should become a starting point for future dialog.

Some key areas of consideration that may affect contracting include:

- Low motivation
- Preoccupation with the consultant selection process
- Questioning of credentials
- Go-betweens
- Over-defining the problem

Contracting is especially useful during the start-up stage of any improvement initiative. Within the OCM, contracting is a useful technique during the clarification and transition planning stages. In particular, the contracting technique should be used in conjunction with forming the Executive Sponsor Team, the Transition Council, and Change Agent teams. It is in these activities that the scope, boundaries, and time frames of team activities are developed. Contracting also is useful whenever there is any unwritten agreement between individuals and the organization. It describes expectations that the organization has of the individual and the individual's expectations of the organization and, therefore, may be advantageous to use in conjunction with schedules and timetables.

Within the boundaries of a change project, contracting is used to convey roles and responsibilities with respect to the Transition Council and Change Agent Teams to develop transition plans and action plans, to clarify project issues, to set the boundaries of consequence management, and to designate feedback interventions. OCM contracting is primarily concerned with the human aspects of a change project and the interaction among the sponsors, agents, targets, and advocates of the change initiative.

Example

Elements of a Contract

A contract has nine essential elements:

1. The boundaries of your analysis.
   Begin with a statement of what problem you are going to focus on.
2. Objectives of the project.
   Identify the organizational improvements you expect if your change project is successful. Three general areas that may need clarification with the initiating sponsor include:
   - To solve a particular technical/business problem
   - To teach the sponsor how to solve the problem for themselves next time
   - To improve how the organization manages its resources, uses its systems, and works internally
3. The kind of information you seek.
   Access to people and information is critical to successful change project implementation. To hedge against ambivalence, be very specific in contracting the type of information you may need, such as technical data, figures, workflows, attitudes of people toward the problem, roles and responsibilities, etc.
4. Your role in the project.
   This is the place to state how you want to work with the sponsor.
5. The product you will deliver.
   Be specific about what you are offering. There ought to be a clear understanding with the sponsor of what your "deliverable" will look like.
6. What support and involvement you need from the sponsor.
   Ensure communication by specifying what you want from the sponsor to make this change project successful.

7. Time schedule.

   Be sure to include any starting time, intermediate milestones, and completion dates for the project.
8. Confidentiality.

   Acknowledge the sponsor's right to choose who sees the results of any report or presentation and who is kept informed of the status of the project.
9. Feedback to you later.

   After the successful completion of the project, you may wish to have the sponsor maintain open communication lines concerning the impact of the project.

Ground Rules for Effective Contracting

- The responsibility for every relationship is 50/50. There are two sides to every story. There must be symmetry or the relationship will collapse. The contract has to be 50/50.
- Contract should be entered into freely.
- You cannot get something for nothing. There must be consideration from both sides, even in a boss/subordinate relationship.
- All wants are legitimate.
- You can say "no" to what others want from you, even sponsors.
- You don't always get everything you want. Both sides may need to make sacrifices in order for agreement.
- You can contract for behavior, but you cannot contract for the other person to change his/her feelings.
- You cannot ask for something the other person does not have.
- You cannot promise something you do not have to deliver.
- You cannot contract with someone who is not in the room, such as sponsors, bosses, and subordinates. You have to meet with them directly to know you have an agreement with them.
- Write down contracts when you can. Most are broken out of neglect, not intent.
- If someone wants to renegotiate a contract in midstream, be grateful that he or she is telling you and not just doing it without a word.
- Contracts require specific time deadlines or duration.
- Good contracts require good faith on both sides. Both sides must truly want to succeed.

Contracting for OCM

- Set up a contracting meeting with the sponsor. This is the first step to effective contracting. The personal interaction during the initial contracting meeting sets the tone for the project. A contracting meeting is normally set up with a phone call. During this call, it is essential to find out the following information:
  - Who requested the meeting?
  - Who will be at the meeting?
  - What will be their roles?
  - How much time is there for the meeting?
  - What is the expected outcome from the meeting?
  - Is a proposal required?
  - What do you want to discuss?
  - Who is the sponsor for this project?

Effective Contracting

The purpose of a contracting meeting is to establish a stable, balanced, and workable contract between the consultant and the sponsor. The following enables effective contracting:

- Personal acknowledgment.
- Communicate understanding of the problems: The consultant needs to communicate an understanding of the problem in ways that acknowledge the unique aspects of the situation, respond to the complexities of the situation, and speak to the sponsor's fear about being beyond help.
- Sponsor wants and offers: "What do you want from me?" There is a difference between what the sponsor wants from the project and what the sponsor wants from you. You need to ask about any constraints about how you should proceed with the project.
- Consultant wants and offers: This is when you, the consultant, put into words what you want from the sponsor to make the project successful. These wants may include such things as enough time to do the job correctly, support from the sponsor at difficult moments, access to the right people and resources, people from the organization to work jointly on the project, agreements on confidentiality, etc.

- Reaching agreement: After exchanging wants with the sponsor, you either reach agreement or get stuck. If you get stuck, it is time to find out what the sponsor does not agree with and what the reservations are. In instances such as this, you may need to take some time to reevaluate these differences.
- Asking for feedback about control and commitment: You need to know how strong the sponsor commitment to the project is.
- Give support: Make supportive statements to the sponsor about his or her willingness to begin this project with you.
- Restate actions: As a final step, make sure you and the sponsor know what each of you is going to do next. After agreeing on the next steps, the contracting meeting is essentially complete.

Ask for Feedback

No contracting meeting should end without asking for feedback from the sponsor. Building commitment starts up front with the first contact with the sponsor. Giving and receiving feedback can be used to start developing the rapport necessary for effective commitment.

Review the Meeting

After the close of the contracting meeting, it is essential to review the meeting before drawing up the contract.

Additional Guidelines for Personal Organizational Relationships

The following guidelines can be applied to personal organizational relationships and are intended to help you successfully contract for what you want in any situation:

- All relationships operate continually; some are just more explicit and/or better negotiated than others.
- All important relationship contracts are expensive; you either pay the price for getting what you want or you pay for not having what you want, but you will pay.
- You cannot have everything you want, but you can have all you are willing and able to pay for.

- The creation of successful contracts requires knowing what you want and committing yourself to paying the price to get it. It also requires that you insist that others perform their role in the process.
- Obligations are not fully clarified until agreements have been reached regarding positive and negative consequences.
- One person can never be more than 50 percent responsible for contract success or failure.
- All contracts are negotiated with insufficient data and, therefore, agreements are always reached with "uninformed optimism."
- Effective contracts can be developed only in situations where all parties have the option of declining unacceptable offers.
- Nothing of real value is ever secured without commensurate payment.
- Do not ask for more than you can pay and do not pay for more than you need.
- Do not ask for something the other party does not control and do not offer anything you do not own.
- Either comply with the terms of the agreement or renegotiate before your obligation is due.
- All aspects of a contract are subject to renegotiations.
- Renegotiation must be completed well before due dates and early enough that if the new proposed stipulations are not accepted, the original obligations can still be met.
- Unilateral changing of terms of a contract is unacceptable.
- Your commitment to fulfill your part of the contract is not just to the other party, but to yourself as well.
- Ultimately, the only truly binding force to sustain a contract is good faith, integrity, morality, professionalism, loyalty, commitment, common goals, and interdependence.

27

Organizational Process Consultation

DEFINITION

Organizational process consultation is the combination of skills in establishing a helpful relationship, in knowing what kind of processes to look for, and in intervening in such a way that processes are improved. Consultants are experts in process design, process redesign, process reengineering, and how processes interrelate with each other making efficient, effective systems. They need to have an actual understanding and know how to apply the software systems that are designed to increase the efficiency and effectiveness of organizational processes.

The goal of organizational process consultation is to influence the value participants place on the relative importance of process versus content in achieving their goals. These individuals can be members of the organization or specialists hired to complete a specific project.

Its Uses

The process consultant works to give group members insight into what is going on between them and other people. Observations are made relative to the various human actions that occur in the normal flow of work, in the conduct of meetings, and formal or informal encounters between members of the group. Of particular relevance is how the behaviors and actions of individual group members impact work. In the past, outside consultants were frequently hired as process consultants. Today, with the rapid changing environment and technology, processes within an organization need to be redesigned frequently and, as a result, organizations are often hiring and training personnel to serve as their internal process consultants. Frequently, Six Sigma Black Belts within the organization are assuming the role of process consultants.

General

"Process consultation is a set of activities on the part of the consultant who helps the client to perceive, understand, and act upon events that occur in the client's environment."* The understanding and analysis of human processes in groups or organizations requires not merely an attitude or a decision to focus on such processes, but also skill and knowledge of what to look for, how to look for it, and how to interpret it. The process consultants are highly skilled individuals with specific knowledge and experience in using methodologies, such as process redesign, process reengineering, Lean, and DMADV (define, measure, analyze, design, and verify).

Preparation

Organizational process consultation should be used throughout the life of a change project, but it is essential to utilize these concepts during the Organizational Change Management (OCM) methodology. It is during these activities that the interaction between members of the teams needs close monitoring and, at times, interventions to ensure that the groups are working effectively.

Organizational Process Consultation Issues

- When to feedback:
  - Feedback to individuals/groups during the process
  - Feedback to individuals/groups after the process
  - Offline coaching
- Structural considerations:
  - Group membership (who is included/excluded)
  - Communication or interaction patterns (past and present)
  - Allocation of work, lines of authority
- Situational variables:
  - Group size
  - Time unit
  - Physical facilities
  - Meeting purpose/topic

* This quote was taken from a powerpoint posted by San Jose State University on OD Interventions Spring 2014. No source was given for the actual quote. www.sjsu.edu/people/harriet.pila/courses/293/

- Group member reaction:
    - Feelings expressed about the observation or observer
    - Response to process observations and recommended interventions

Organizational Process Consultation Interventions

- Agenda setting interventions:
    - Agenda items that direct attention to process issues
    - Process analysis periods during or at the end of each meeting
    - Meetings devoted strictly to process (i.e., workshops, retreats)
- Task-related interventions:
    - Establish common goal
    - Seek clarification of roles
    - Acknowledge need for procedures
- Relationship-related interventions:
    - Inquire how candid and open people feel they can be
    - Ask how control, status, and power are getting established
    - Acknowledge group norms
    - Recognize the acceptance of differences
    - Establish ground rules

Task-Oriented Behaviors

- **Initiating:** The beginning of an activity or plan. There are two types of initiation. Both involve making sure some sort of suggestion, often independent of any specific request from the group.
- **Process initiation** relates to how a meeting or issue is to be approached, proposing a new approach to problem resolution, or some new procedure.
- **Content initiation** relates to new ideas, problems, issues, or direction; suggesting a new or alternative goal, task, or solution.
- **Building:** Building expands or develops others' suggestions or ideas.
- **Giving information:** This type of behavior offers facts, generalizations, or information.
- **Seeking information:** This type of behavior is represented by requesting examples, illustrations, elaborations, suggestions, or ideas from others.
- **Clarifying/testing understanding:** Clearing up confusion or misunderstanding; interpreting meaning.

- **Gatekeeping:** Asking someone to hold off introducing a new topic while another is being discussed; intervening to prevent people from talking at the same time; monitoring time.
- **Summarizing:** Pulling together related ideas restating suggestions after the group has discussed them; providing closure.
- **Agreeing:** Expressing agreement or support of another's ideas.
- **Disagreeing:** Expressing difference of opinion or contrasting views of another's opinion.
- **Recordkeeping:** Documenting minutes and action items.
- **Evaluating/Following up:** Providing rigorous and regular evaluation of performance and taking appropriate action for assigned tasks.

Relationship-Oriented Behaviors

- **Trusting:** Establishing an environment where absolute trust and open, candid communication across departments or status and power levels are the norm.
- **Listening:** Helping to keep communication channels open through attentive, sincerely interested listening, sometimes over long periods.
- **Encouraging:** Listening attentively to other's ideas; reinforcing the ideas verbally or nonverbally; encouraging participation from silent members; being friendly, warm, and responsive to others and providing extensive approving and sympathetic encouragement.
- **Accepting:** When your own idea or status is involved in a conflict, offering a willingness for complete, active, and genuine acceptance of other's motives.
- **Harmonizing:** Trying to reconcile disagreements by getting people to explore the causes behind their differences rather than just arguing over opposing feelings or concerns; asking for verification.
- **Expressing feelings:** Sensing feelings, moods, relationships within the group; sharing own feelings or concerns; asking for verification.
- **Advising:** Testing whether the individual or group is satisfied with its procedures and progress, and providing a considerable amount of helpful input.

Example

Conducting Organizational Process Consultation

Organizational process consultation involves stimulating the group's interest in working their process issues by asking the group or individuals in the group questions that direct their attention to the issues that are perceived to be interfering with effective group performance. Organizational process consultation also helps the group learn to gather its own data and draw its own conclusions.

Consultant's Role

As an outsider, it is absolutely essential that he/she be seen as an advisor and helper to the individuals that are involved in the process being improved and the process owner. This often results in the design phase moving a little slower as he/she helps the team realize the large number of improvement opportunities that are available to them. This results in building a sense of ownership in the new process design that decreases the time required to implement the new process and greatly impacts the overall positive performance of the new process.

- Observe the group's process.

  In all human interactions there are two major ingredients: content and process. The first deals with the subject matter upon which the group is working. In most interactions, the focus of attention of all persons is on the content. The second ingredient, process, is concerned with what is happening between and to group members while the group is working. Group processes, or dynamics, deal with such items like morale, feeling, tone, atmosphere, influence, participation, styles of influence, leadership struggles, conflict, competition, cooperation, etc. In most interactions, very little attention is paid to process, even when it is the major cause of ineffective group action. Sensitivity to group process better enables one to diagnose group problems early and deal with them more effectively. Because these processes are present in all groups, awareness of them enhances a group member's worth to a group and enables him or her to be a more effective group participant.
- Monitor the process: When you see one person taking control over the group, people not listening, people dropping out, or any activity that slows or interferes with the team's process, record the behavior mentally or take notes.
- Provide feedback process observations to an individual, group, or organization.

  There are several ways to provide feedback process observations (i.e., during the meeting, after the meeting, or online). One of the most effective means of dealing with process issues is for the process observer to give his or her comments at the end of the meeting if it was contracted that all members get feedback. Often the consultant follows up with the leader via offline coaching and gives some specific feedback on how his/her behavior influenced the group's process. This can then lead to discussion.

- At the end of the meeting, allow time for a process discussion.

  Simply ask, "How did the group do today?" Solicit responses from the group. Ask the group to discuss causes for conflict, delays, or any behaviors that have interfered with group effectiveness. Stay in tune to what the issues are. Don't allow people to blame or point to who is an issue. All members are equally responsible for what occurs during the process as long as they are present. Eventually, members will realize that they are going to be discussing the process and this will increase their awareness of the process issues that occur during the regular task part of the meeting.
- Prescribe the appropriate intervention.

  With the group or group leader, agree on the sequence or set of interventions needed to increase effectiveness (i.e., team building, goal setting, role clarification).
- Follow up and monitor intervention effectiveness.

  Continue to observe the process for effectiveness of intervention. Recontract with the individual or group regarding your role and purpose.

Observer Role

A process "observer" does not participate in the meeting. His or her job is to observe how the team is functioning and to give feedback to the members at the end of the meeting or during the meeting if requested or contracted for prior to the meeting.

The observer watches the team interaction and the contributions of the individual members. At the end of the meeting, the observer gives his or her comments to the group for acceptance or discussion. Remember that perceptions of what happens in a group vary among members, so different views are to be expected. If the observer offers his/her views as his/her perception, not as fact, a team can discuss the offering freely. The important point is that the team talks about its process and that the team members have a chance to alter their behavior if doing so improves the performance of the team.

It is important to have some understanding of the culture the individual or group is operating in as this influences behaviors and norms. What may be appropriate in one group or organization may not be in another.

The group defines essentially what the effective means are for them (i.e., establishing a strategy, achieving a goal, making a decision). The organizational process consultant must not impose his/her interpretation on the individual or person.

A typical example of the improvements that can be brought about by focusing on an organization's processes is the result Federal-Mogul Corporation (Southfield, MI) had in its new product development process. The company reduced its development process cycle time from 20 weeks to 20 business days, resulting in a 75 percent reduction in throughput time (Harrington, 2011).

Software

See the software listed in Simulation Modeling and Flowcharting chapters.

REFERENCE

Harrington, H. J. 2011. *Streamlined process improvement*. New York: McGraw-Hill, p. 323.

28

Organizational Process Improvement/ Business Process Improvement

DEFINITION

Organizational Process Improvement (OPI) is a combination of two methodologies: process reengineering and process redesign. OPI is often called Business Process Improvement (BPI). OPI is a systematic approach to bringing about step-function improvement in an organization's processes. It focuses on increasing adaptability, efficiency, and effectiveness while reducing cost and cycle time.

Its Uses

OPI is designed to reduce costs and improve customer satisfaction by streamlining the operating processes within an organization. Typically it will result in reducing cycle times and costs between 30 to 90 percent.

General

OPI consists of two methodologies: process reengineering and process redesign.

- The process redesign methodology focuses on streamlining the present process, making it more efficient, effective, and adaptable. Improvements related to process redesign typically vary between 30 to 60 percent reduction in costs and cycle time.
- The process reengineering methodology focuses on creating a completely new innovative process and relies heavily on information

technology to support the improvement activities. Improvements related to process reengineering typically vary between 60 to 90 percent reduction in costs and cycle time. The risks related to process reengineering approaches are much higher and the future state solution usually takes a longer time to define and it is more costly to implement than the solutions defined by the process redesign methodology.

Preparation for Process Redesign Methodology

The process redesign approach focuses the efforts of the Process Improvement Team (PIT) on refining the present process. Process redesign is normally applied to processes that are working fairly well today. Typically, process redesign projects will reduce cost, cycle time, and error rates between 30 and 60 percent. With process redesign, it takes between 80 and 100 days to define the Best-Value Future-State Solution (BFSS). This is the correct approach to use with approximately 70 to 90 percent of major business processes. This approach is used if improving the process's performance by 30 to 60 percent would give the organization a competitive advantage.

In redesigning processes, an as-is simulation model is constructed. Then, the following streamlining tools are applied:

- Bureaucracy elimination
- Value-added analysis
- Duplication elimination
- Simplification methods
- Cycle time reduction
- Error proofing (current problem analysis)
- Process upgrading (organizational restructuring)
- Simple language
- Standardization
- Supplier partnerships
- Automation, mechanization, and information technology

You will note that the information technology enablers are applied after the present process's activities have been optimized. Once the process's activities have been optimized, information technology and computerization best practices are used to support the optimum process. This truly puts information technology in the role of being a process enabler rather than a process driver. With the redesign concept, the PIT does not create new information technology (IT) applications, but takes advantage of the best practices that are already proved. Often, a process comparative analysis is conducted in parallel to the redesign activities to ensure that the redesigned process will be equivalent to or better than today's best practices.

- The five phases of process redesign.

  The complexity of our business environment and the many organizations involved in the critical business processes make it necessary to develop a very formal approach to process redesign. This methodology is conveniently divided into five subprocesses called *phases* that consist of a total of 28 activities.
  - Phase I: Organizing for Improvement (8 activities)
  - Phase II: Understanding the Process (6 activities)
  - Phase IIA: Conducting a Comparative Analysis (Optional)
  - Phase III: Streamlining the Process (6 activities)
  - Phase IV: Implementation, Measurements, and Control (5 activities)
  - Phase V: Continuous Improvement (3 activities)

Preparation for Process Reengineering Methodology

Process reengineering is the most radical of the OPI approaches. It is sometimes called *process innovation* because its success relies heavily on the PIT's innovation and creative abilities. Process reengineering is also called *Big Picture Analysis* or *New Process Design*. We like the term New Process Design best because it uses the same approach that would have been used if the organization were designing the process for the first time.

Process reengineering, when applied correctly, reduces cost and cycle time between 60 and 90 percent and error rates between 40 and 70 percent. It is a very useful tool when the current-state process is so out of date that it is not worth salvaging or even influencing the BFSS. Process reengineering is the correct answer for 5 to 20 percent of the major processes within an organization. If you find it advantageous to use process reengineering in more than 20 percent of your major processes, the organization should be very concerned, as it may be indicative of a major problem with the way the organization is managed. This management problem should be addressed first, before a great deal of effort is devoted to improving processes that will not be maintained.

The process reengineering approach to OPI allows the PIT to develop a process that is as close to ideal as possible. The PIT steps back and looks at the process with a fresh set of eyes, asking itself how it would design this process if it had no restrictions. The approach takes advantage of the available process enablers, including the latest mechanization, automation, and information technology techniques, and improves upon them to come up with new IT products. Process reengineering challenges all of the paradigms that the present process is built upon. This process stimulates the PIT to come up with a radical new process design that is truly a major breakthrough.

The process reengineering approach provides the biggest improvement, but it is the most costly and time-consuming OPI approach. It also has the highest degree of risk associated with it. Often, the process reengineering approach includes organizational restructuring, and that can be very disruptive to the organization. Most organizations can only effectively implement one change of this magnitude at a time.

Process reengineering consists of four phases (Figure 28.1). They include:

- Phase I: Organizing for improvement
- Phase II: Designing the best-value future-state solution (BFSS)
- Phase III: Implementation
- Phase IV: Continuous improvement

Phase I: Organizing for Improvement

An Executive Improvement Team (EIT) is formed. Process owners and Process Improvement Teams (PITs) are assigned, process boundaries are defined, total process measurements are developed, and initial business process improvement project plans are developed and approved.

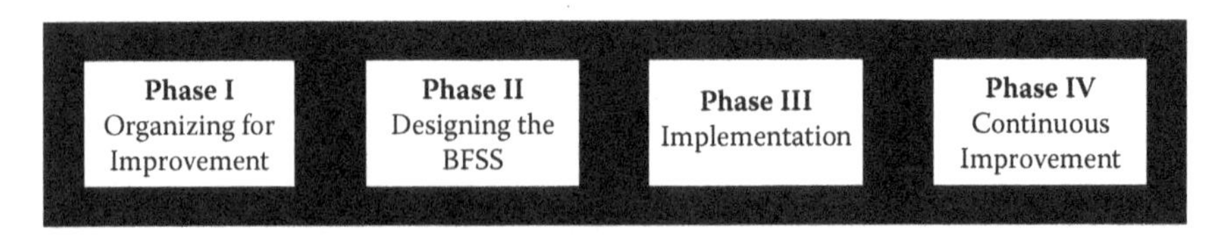

FIGURE 28.1
The four phases of process reengineering.

Phase II: Designing the Best-Value Future-State Solution (BFSS)

The process reengineering Phase II approach to developing a BFSS consists of six tasks:

Task 1: Big picture analysis
Task 2: Theory of ones
Task 3: Process simulation
Task 4: Alternative future-state solutions
Task 5: Process modeling
Task 6: Preliminary implementation plan approval

Task 1: Big Picture Analysis

During this task, the PIT is not constrained in the development of a vision document that defines the ideal process. The only restriction is that the results of the process reengineering activities must be in line with the corporate mission and strategy. They also should reinforce the organization's core capabilities and competencies. All other paradigms can and should be challenged. Before the PIT starts to design the new process, it needs to understand where the organization is going, how the process being evaluated supports future business needs, and what changes would provide the organization with the most important competitive advantage.

Once this is understood, the PIT can develop a vision statement of what the best process would look like and how it would function. In developing the vision statement, the PIT needs to think outside the normal routine (think outside the box) and challenge all assumptions, challenge all constraints, question the obvious, identify the technologies and organizational structures that are limiting the process, and define how they need to be improved to create a process that is better than today's best. The vision statement defines only what must be done, not what is being done. Usually, the vision statement is between 10 and 30 pages and, in reality, is more like a new process specification. It will define all of the process, information technology, organizational and people enablers that would be applied in designing the new process.

Task 2: Theory of Ones

Once the vision statement is complete, the PIT should define what must be done within the process from input to delivery to the customer. It needs to question why the process cannot be done using only one resource unit (person, time, money, space, etc.). The PIT should be a miser in adding activities and resources to the process (Figure 28.2).

To use the theory of ones, the PIT sets the minimum quantity of units that it is trying to optimize. For example, if the PIT is interested in optimizing cycle time and the previous cycle time was five days, it might ask the question, "What if I had to do it in one second? What enablers would have to be used, and what paradigms would have to be discarded to accomplish this?" Basically, four sets of enablers are addressed:

1. Process enablers
2. Information technology enablers
3. People enablers
4. Organizational enablers

After the PIT has looked at each of the enablers and challenged each paradigm, a process design is defined to accomplish the desired function. The resulting process is compared to the vision statement from Task 1. If the PIT gets an acceptable answer, it goes forward. If not, it repeats the cycle with a new objective of doing the total process in one minute. At some point in time, the new process design and vision statement will be in harmony. As you can see, reengineering is very much an iterative process.

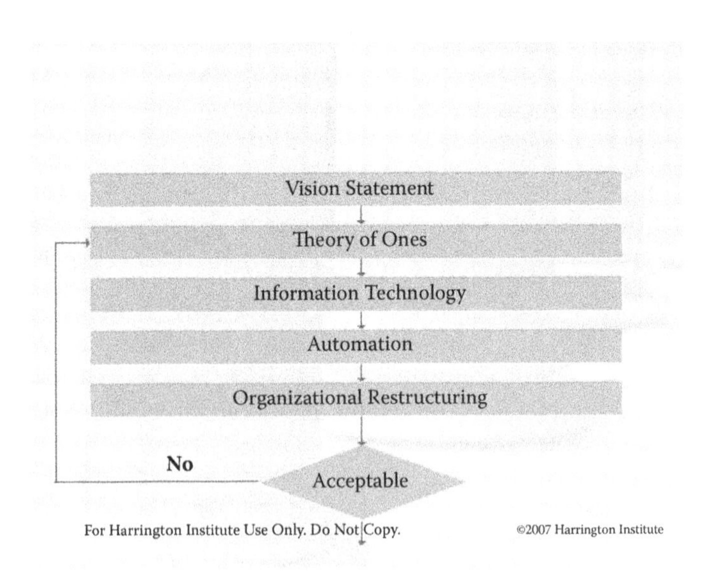

FIGURE 28.2
Theory of ones.

Task 3: Process Simulation

When the new process design is theoretically in line with the objectives set forth in the vision statement, a simulation model is constructed. The simulation model is then exercised to evaluate how the new process design will function.

Task 4: Alternative Future-State Solutions

We recommend that a minimum of three future-state solutions be developed. This allows you to have some options in making your final process selection. It is usually helpful in developing the other future-state solutions to set different objectives for each solution.

Task 5: Process Modeling

Once the BFSS is selected, the theoretical model is physically modeled to prove the concepts.

Task 6: Preliminary Implementation Plan Approval

The PIT will now prepare a preliminary implementation plan and associated budget. This allows the PIT to document in greater detail the BFSS that it developed and explain what considerations were included in the return-on-investment estimates. The PIT should submit the preliminary implementation plan to the executive committee along with recommendations related to who should be assigned to implement the new process and its accompanying measurement system. If the executive team approves the preliminary implementation plan and the associated budget for the project, the project is ready to move into Phase III.

Phase III: Implementation, Measurements, and Control

Increased emphasis is placed on Organizational Change Management during Phase III. The BFSS solution is phased in with the appropriate number of trial runs that verify the magnitude and impact of each change. The simulation model is updated to reflect these changes.

Phase IV: Continuous Improvement

Process improvement plans are reviewed by the EIT and approved. The process evolves through a series of six qualification levels. Each time a change to the process is implemented, the simulation model is updated. The outputs from Phase IV include:

1. The EIT takes over the responsibility of continuous improvement for their part of the process.
2. Process changes are implemented.
3. Process improvement results are measured.
4. Process simulation model is updated.

Example

Typical examples of improvements realized when the process redesign approach was used are shown in Table 28.1.

Software

There are currently a number of software products on the market that can assist you in creating process flowcharts. Some of them are generic tools, such as drawing programs, while others have been created specifically for flowcharting. One product we suggest is a software package produced by Edge Software Inc. called WorkDraw. This is an advanced process

TABLE 28.1

Examples of Improvements Using Process Redesign Approach

IBM's Communication Network	12 month results:
33 Countries	3X improvement in response time
	Availability up from 86 to 95 percent
NutraSweet	20 percent decrease in order management costs
(Sales, marketing, and logistics)	20 percent cost reduction
	20 percent increase in customer service
Colgate	25 percent reduction of distribution and order management costs
Palmolive	Increased sales based on customer service improvements
(Sales, marketing, and logistics)	
Toyota	Market share increase of 40 percent
(Sales, marketing, and logistics)	Profit enhancement in excess of 30 percent
	Administrative process error reduction rate reduced to less than 5 percent

modeling application that includes flowcharting capabilities. Typical other software packages include:

- Visio 14.0.6
- SmartDraw VP 19.1.3.2®
- Flowcharter 14.1.2
- Edraw Flowchart 6
- EDGE Diagrammer 6.24
- WizFlow Flowcharter 6.24
- RFFlow 5.06

ADDITIONAL READING

Cooper, R., and R. Slagmulder. 1997. *Target costing and value engineering.* Portland, OR: Productivity Press.

Harrington, H. J. 1991. *Business process improvement.* New York: McGraw-Hill.

Harrington, H. J. 1997. *The business process improvement workbook.* New York: McGraw-Hill.

Harrington, H. J., and J. S. Harrington. 1995. *Total improvement management.* New York: McGraw-Hill.

Harrington, H. J., and K. Tumay. 2000. *Simulation modeling methods.* New York: McGraw-Hill.

29

Pareto Analysis

DEFINITION

The Pareto diagram is a specialized type of column graph that is created to simplify comparisons between items.

Its Uses

Pareto analysis allows the user to make comparisons between a number of problems and a number of causes. The analysis itself is based on the Pareto Principle, or "80–20" rule, which states 20 percent of contributors to a problem are responsible for 80 percent of the problem.

General

This special type of column graph helps the user to determine which problems one should solve and in what order. The largest problem or most important cause is identified by ranking all items according to size. The result is a cumulative percentage graph that helps to establish priorities in problem solving by displaying the proportionate importance of certain categories.

Preparation

Using Pareto Analysis

- Gather data on problems to be investigated.
- Identify what is to investigate.
- Collect data about each item.

- Prepare a data table.
- List each item on a worksheet and show the number of occurrences of each item. If several minor item categories exist, an "Other" category may be created to represent multiple infrequent problems.
- Arrange the data from largest to smallest and total the column. Calculate for each category the percentage of the total items occurring. Also, for ease of constructing the Pareto diagram, total the cumulative percentages in another column in the worksheet (Table 29.1).
- Create a bar graph depicting the data.
- Draw one horizontal and two vertical axes. Divide the horizontal axis into equal partitions, one for each item.
- List the items across the horizontal axis in decreasing order of occurrence. The item that occurs most often should be placed on the far left. If an "Other" category has been created, it should be placed on the far right of the axis.
- Scale the left-hand vertical axis such that the top of the axis is a value equal to that of the total number of all item occurrences.
- Scale the right-hand vertical axis with a percentage scale ranging from 1 to 100 percent. The 100 percent value must be directly across from the total on the top of the left-hand axis. The percent scale on the right-hand axis is normally in increments of 10 percent.
- Create frequency columns.
- Construct a frequency column for each problem category listed on the horizontal axis.
- Plot the cumulative values.

TABLE 29.1

Pareto Analysis: Data Table

Item Name	Number	% of Total	Cum. %
High Amperage	26	40.0	40.0
Wrong Color	12	18.5	58.5
Low Amperage	9	13.8	72.3
Too Small	7	10.8	83.1
Too Large	5	7.7	90.8
Low Voltage	4	6.2	96.9
High Voltage	2	3.1	100.0
Total	65	100.0	

- Place a dot in line with the upper right corner of each column, at a height corresponding to the number in the cumulative percentage column in the worksheet.
- Beginning with the lower left corner of the first column on the diagram, connect the dots up to the 100 percent point on the right-hand vertical axis.
- Label the graph.
- Label each axis and add a legend. The legend includes the source of the data, the date prepared, where the data was collected, who collected it, the period covered, and any other pertinent information.

Example

See Figure 29.1.

Software

Some commercial software available includes, but is not limited to:

- Excel®
- Minitab®
- JMP®

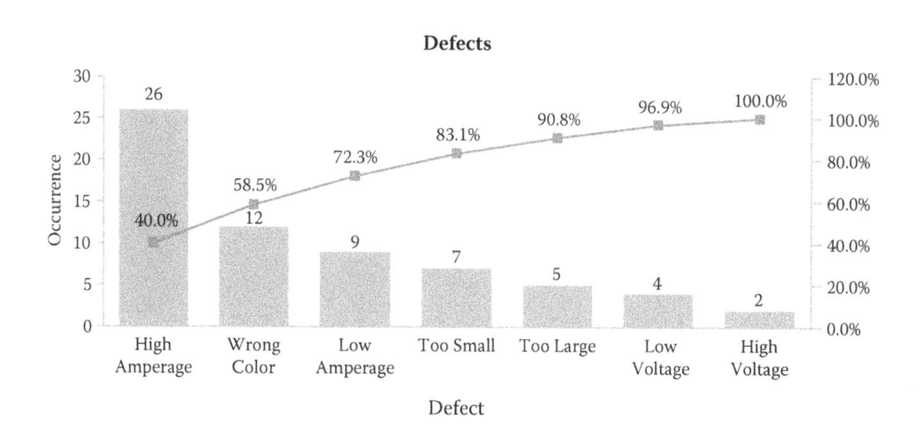

FIGURE 29.1
Typical Pareto analysis.

30

Prioritization Matrices

DEFINITION

A prioritization matrix is a narrowing technique that is used to rank large lists of alternatives.

Its Uses

Prioritization matrices are used to evaluate large lists of alternatives.

General

This narrowing technique also can be used to select certain items from a list of alternatives. Items are generally prioritized mathematically by comparing all items to each other. Criteria are developed for comparison purposes and each item is prioritized based on the established criteria.

Preparation

Evaluating Using Prioritization Matrices

- Determine the alternatives to be evaluated.

  The group must agree on the relevance of the chosen items. If the group is prioritizing evaluation criteria, the criteria should be worded in terms of the ideal result. Wording should not be neutral. For example, a criterion could be "Easy to Implement," but should not be "Ease of Implementation."

- Develop a matrix.
  Develop a matrix to determine the desirability of the alternatives by comparing the items against each other. List the alternatives across the top of the horizontal axis and down the vertical axis, using the same listing order for both axes. Form the matrix by drawing horizontal and vertical grid lines. Cross out the cells that correspond to the same item on the horizontal and vertical axes (the crossed out cells should form a diagonal line from the upper left to lower right of the matrix). Devise a numeric scale for comparing alternatives. The scale might look like the following:
  1 = Equally desirable
  5 = More desirable
  10 = Much more desirable
  1/5 (0.2) = Less desirable
  1/10 (0.1) = Much less desirable
  Any form of numeric representation can be used to prioritize.
- Compare all vertical axes alternatives against the horizontal axes alternatives.
- Place the appropriate numeric value in each cell.
- Total the columns and rows to determine which items are most important.

TABLE 30.1

Typical Prioritization Matrix

	Low Cost to Implement	No Customized Technology	Quick to Implement	Easily Accepted by Users	Minimal Impact on Other Departments	Total Across Rows (% of Grand Total)
Low Cost to Implement		5	1/10	1/10	1/5	5.5 (8%)
No Customized Technology	1/5		1/5	1/10	1/5	0.7 (2%)
Quick to Implement	10	5		1/10	1/5	15.3 (21%)
Easily Accepted by Users	10	10	10		1/5	30.2 (40%)
Minimal Impact on Other Departments	5	5	5	5		20.0 (28%)
Column Total	25.2	25.0	15.3	7.3	0.8	73.6 (100%)

Example

See Table 30.1.

Software

Some commercial software available includes, but is not limited to:

- Excel®

31

Process Capability Analysis (Cp)

DEFINITION

Process Capability Analysis (Cp) is a statistical comparison of a measurement pattern or distribution to specification limits to determine if a process can consistently deliver products within the specification limits. It is a measure of the relationship between common system variation and the specification limit.

Its Uses

Process Capability Analysis can be used on the entire subprocess or any of its activities. The goal is to identify those activities within the process that will not meet the product or service requirements. This technique also can be used to determine how efficient and effective a proposed process is relative to its capability of performing.

General

Process Capability Analysis evaluates a system with respect to its specified tolerance. The analysis can determine whether the currently designed, equipped, and operated system can meet its requirements; if it cannot, the analysis identifies the proportion of the output that is expected to be defective. The analysis assumes that the subprocess or activity under evaluation is normally distributed with 99.73 percent of the observations within three standard deviations of either side of the mean (six standard deviations in total). The analysis also assumes that the process is stable under a state of statistical control. If the process being studied does not adhere to both of these assumptions, Cp should not be used.

Preparation

Analyzing Process Capability

Process capability studies provide input for making the appropriate management decisions and establishing priorities for which subprocesses or activities have opportunities for improvement. Every situation must be weighed on the circumstances, the cost of sorting output versus the cost of making system improvements to reduce the variation from common causes. The goal is a never-ending quest for improving process performance.

Collect the Data

- Collect the data after becoming familiar with the process.
  This approach generically describes "process" characteristics; these may apply to both subprocesses and activities.
- Determine the size and sequence of subgroups to be measured.
  A subgroup size of four to five samples is usually adequate and should represent as nearly as possible subprocess performance under one set of conditions. Refer to Data Gathering by Samples and Surveys (Chapters 13 and 14) technique for additional guidance on sampling.
- Decide the number of subgroups to be measured and the frequency of the subgroup collection.
  The goal is to capture the total variation that exists in the subprocess or activity by choosing a collection frequency that will detect all changes the subprocess or activity will normally undergo.

Verify That the Process Data Is Normally Distributed

- Verify the distribution.
  This uses a graphic plot to develop a histogram that can be analyzed to determine if its shape fits the requirements to be called a normal distribution. If the process data are not normally distributed, process capability analysis cannot be conducted.

Calculate the Mean and Range

- Calculate the mean and range for each subgroup as well as the average mean and range for the total of all subgroups.

Develop the Control Chart for the Process

- Chart the process.

  The population mean will be the center line for the control chart. Refer to the Control Charts technique (Chapter 9) for a detailed explanation on the construction of control charts.

Evaluate the Control Chart

- Verify that the process is in a state of statistical control.

  If the process is not stable and predictable, it must be brought under control before continuing Process Capability Analysis.

Determine the Specification Limits for the Process under Observation

- The specification limit is a process boundary established independently of the process and is based primarily on the requirements of the product or service. The specification limit will have an upper specification limit value (USL) and a lower specification limit value (LSL). The tolerance is the width between the lower and upper specification limits.

  Because the specification limits should be determined independent of the data observations gathered for the subgroup, the tolerance does not have to directly correspond to plus or minus three standard deviations about the mean. Figure 31.1 is an example of the specification limits and tolerance compared to a normal distribution. As illustrated, observations can fall within the six standard deviation limit, but still be intolerable, depending on where the tolerance limit is set.

  There is a major difference between the limits on a control chart and the limits in Process Capability Analysis. Control chart limits are statistically calculated. Limits in Process Capability Analysis are specifications based on product requirements. The two sets of limits are mutually exclusive; therefore, a process can be statistically in control, but still be producing outputs that do not meet specifications.

Example

See Figure 31.1.

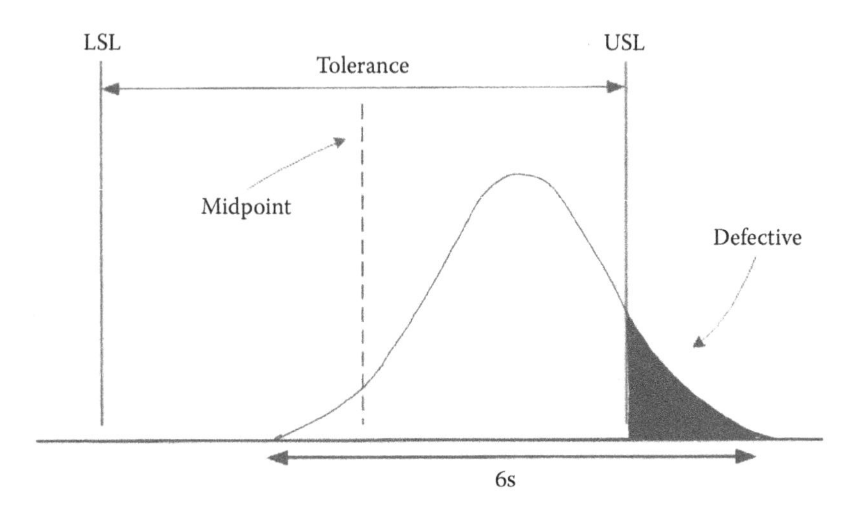

FIGURE 31.1
Example of the specification limits and tolerance compared to a normal distribution.

Analyze the Data for Process Capability

- Calculate the standard deviation of the entire process. This can be adequately approximated using the estimated standard deviation formula: s = R/d2, where R is the average of the subgroup ranges and d2 is a control chart constant that is dependent on the sample size (d2 is provided in the control chart technique).
- Calculate the location of the process in Z units relative to its upper and lower specification limits (Zu and ZI) using the formulas Zu = (USL – X)/s and Z1 = (X – LSL)/s.

  These values determine the number of standard deviations away from the mean that the upper specification and lower specification limits are set.
- Calculate the upper and lower process capability indices (CPU and CPL) using the Z values and the formulas: CPU = Zu/3 and CPL = Zl/3.

  Because, ideally, there should be at least three standard deviations between both the upper and lower specification limits and the mean, CPU and CPL should both be greater than or equal to one. If either of these values are less than one, the process is currently capable of producing outputs that are outside of the product specifications.

Perform Further Analysis If Necessary

A process can be characterized in two ways: inherently capable and operationally capable. A process is inherently capable when, under a state of statistical control, the six standard deviation variation of its frequency distribution is no wider than the tolerance. A process is operationally capable when, under a state of statistical control, the six standard deviation variation of its frequency distribution is no wider than the tolerance and the six standard deviation variation falls within the tolerance range.

Further Analysis (2)

Process outputs may fall outside the tolerance limit, but still have a six standard deviation variance that is less than the tolerance. A process operating under this condition is considered off-centered. In this situation the process still may be capable of producing outputs within specifications; all that is required is to center the process. It should be noted that, while centering a process is a fairly simple task, reducing the total process variation by narrowing the six standard deviation distribution is a complex and time-consuming activity.

The inherent capability of the process is designated by the Cp index and can be calculated using the formula Cp = Tolerance/6s. Therefore, if the tolerance width is exactly the same as the six standard deviation width, the Cp index is equal to 1.0. Because the six standard deviation limit only encompasses 99.73 percent of the output, even if the process is operated at the midpoint of the tolerance, a Cp = 1.33 (t4s) is usually used as a guideline before the process is considered capable. Table 31.1 provides guidance in interpreting the Cp index.

TABLE 31.1

Cp Decision Table

Cp Value	Decision
Greater than 1.3	Process is capable
Between 1.00 and 1.33	Process is capable, but should be monitored as Cp approaches 1.0
Equal to 1.00	Capability limit has been reached and process is producing outputs equal to the tolerance
Less than 1.00	Process is not capable

Software

Some commercial software available includes, but is not limited to:

- Minitab®
- JMP®
- Excel®
- QI Macros™
- Sigma Excel®

ADDITIONAL READING

Bothe, D. R. 1997. *Measuring process capability: Techniques and calculations for quality and manufacturing engineers*. Milwaukee, WI: ASQ Quality Press.

Burr, I. W. 1976. *Statistical quality control methods*. New York: Marcel Dekker.

Deming, W. E. 1986. *Out of the crisis*. Cambridge, MA: MIT Center for Advanced Engineering Study.

Ishikawa, K. 1982. *Guide to quality control*, 2nd ed. Tokyo: Asian Productivity Organization.

Montgomery, D. C. 1985. *Introduction to statistical quality control*. New York: John Wiley & Sons.

Orr, E. R. 1975. *Process quality control: Trouble-shooting and interpretation of data*. New York: McGraw-Hill.

Shewhart, W. A. 1931. *Economic control of quality manufactured product*. New York: D. Van Nostrand Company, reprinted by ASQC.

Wadsworth, H. K., et al. 1986. *Modern methods for quality control and improvement*. New York: John Wiley & Sons.

32

Project Charter

DEFINITION

A project charter is a document that formally organizes the project, thereby authorizing the project leader to enlist organizational resources to accomplish its objectives. It also defines what the project is responsible for accomplishing.

Its Uses

A project charter is the cornerstone of a Lean/Six Sigma project, and, if it is used properly, can be an important tool for managing the expectations of the project sponsor and other stakeholders. The charter is the foundation of any project. The project scope and assumptions stated in the project charter are updated as they change during the course of the project.

General

Development of a project charter is as important as the charter itself. If the chartering process has been poorly done and if the project sponsor and project manager have not truly agreed on its contents, the project charter is worthless.

A project charter is essentially a contract or job description based on results. It may not be legally binding on the parties involved, but it represents a formal agreement and commitment between the project manager and the project sponsor. It is the project manager's professional responsibility to treat this agreement seriously and make every effort to meet the commitment it represents.

The project charter and the Lean/Six Sigma project management approach are designed to help the project manager:

- Document the expected outcome between the project sponsor and the project manager
- Provide a clear statement of the purpose of project and what the team is committed to deliver
- Define the project roles and responsibilities
- Define the development process and approach that will be used to manage the project
- Establish the ground rules for the project
- Provide a baseline for scope and expectation management

Preparation

Developing a project charter:

- Initiate the chartering process:
  - The initial project charter and the project charter
- Chartering a planning or definition project
- Creating project approaches, plans, and procedures
- Negotiate and agree on the project charter details
  - Estimating is not negotiating
- Obtain approval of the project charter

Example

1. Developing a project charter:
   a. The project sponsor must be confirmed and the necessary authority to proceed obtained.
   b. The chartering process can begin in a number of ways. You will usually find yourself in one of two situations:
      i. You have to develop initial project charters for future projects at the end of a current project (Stage N-1 of the life cycle currently ending).
      ii. You have to create a final project charter from an inherited initial project charter (Stage 1 of the new project about to begin).

In both these circumstances, the chartering process has already been initiated. In other circumstances you may have to respond to a more tactical or opportunity/problem-driven user.

Initiation of a new charter has to be handled very carefully to establish a sound baseline for the project. Project chartering takes on the characteristics of a project justification and proposal process.

Confirm with the project sponsor and establish the required Lean/Six Sigma baselines. If the baselines do not exist, develop them prior to, or in parallel with, the chartering process.

2. The initial project charter and the project charter.

   Project chartering is normally a two-step process. An initial project charter is developed in stage N-1 of a project. For example, during the development planning stage, an initial project charter is created for every subsequent development project.

   In stage 1 of these projects, the initial project charter is refined to produce a final project charter and a detailed project plan. A high-level summary of the detailed project plan is included in the project charter for approval purposes (Figure 32.1).

3. Chartering a planning or definition project.

   Because an initial project charter will not normally exist for planning or definition of projects, these roadmaps both contain all the work required to develop a project charter from a zero baseline.

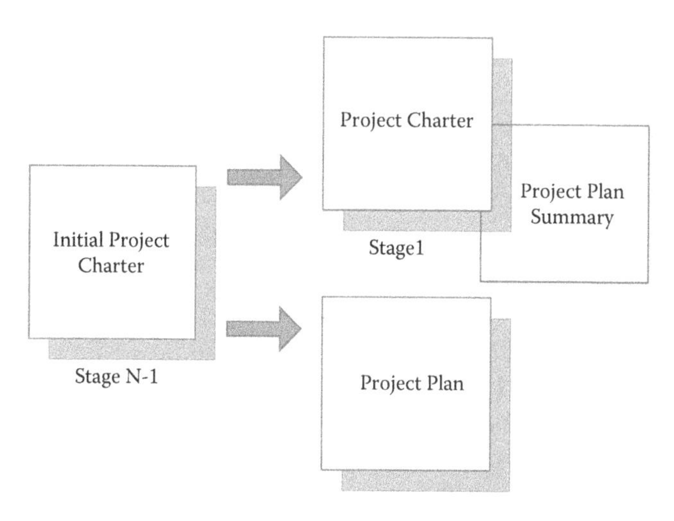

FIGURE 32.1
The initial project charter and the project charter.

4. Pay attention to the sponsor relationship.

   This process should improve your working relationship with the project sponsor. If this is the first time you have met the project sponsor, remember that the way you interact during chartering will often set expectations throughout the project:

   a. You should try to create an environment of openness in mutual trust.
   b. Do not avoid any issues that arise.
   c. Be formal in your handling of issues and assumptions and listen to what the sponsor wants to get out of the project.

   Listening skills are critical, but also pay attention to nonverbal communication.

5. Formulate the goals, objectives, and opportunity/problems.

   Formalize and document the goals, objectives, opportunities, and problems. Determine why the project is taking place.

   Sometimes when the chartering process starts, the project sponsor will not have a clear formulation of the problem. During the first stages of chartering, work together to define the goals, objectives, and the opportunity/problems that initiated the process. Discuss how the resultant activities will address and mitigate or remove the problem. This information will become the project goals and objectives and the project management approach in the project charter:

   a. Chartering is an interactive activity.

      Chartering, like the development work in the project, is highly interactive and demands a high level of user involvement:

      i. Meet with the project and executive sponsors frequently.
      ii. Ensure that the project charter is accurate.
      iii. Respond quickly to their feedback.

   When you feel you have the opportunity/problem formalized, meet with the project sponsor to ensure that you are in agreement about what the project is trying to accomplish.

6. Develop the project charter components:

   a. Develop the project goals and objectives.

      The project goals and objectives should align with the goals and objectives of the enterprise. Ensure that the project objectives are clearly defined and measurable.

   b. Develop the project scope.

      The project scope is the most important part of the charter document. It describes exactly what will be produced, how long it will take to produce it, and how much the enterprise will pay to have it produced. It also will define what will not be addressed during the project. Keep the project sponsor involved in the development of the project scope.

c. Develop the project organization and the project management approach.

   Ensure buy-in of the project management approach, specifically the approach to issue management and scope management.

d. Develop the project plan summary.

   A more detailed project work plan will be developed by the project manager, but normally only a high-level summary should appear in the project charter unless the project sponsor wants to see more detail.

7. Creating project approaches, plans, and procedures.

   For most of the project control activities (issue management, scope management, knowledge coordination, and risk management), Lean/Six Sigma suggests the creation of an approach, a plan, and procedures. Each of these documents serves a different purpose in terms of its audience and its content.

   A project approach is designed to inform the project sponsor that you are aware of the need for the activity and are prepared to manage it. For this reason, an approach typically contains a description of the problem, i.e.:

   a. The need to track and resolve issues
   b. Why tracking changes to the project baseline is important
   c. An overview of how you propose to manage the problem

   A plan is designed to integrate the approach into the work plan of the project and to formalize who will be expected to participate in the activities. Therefore, a plan contains:

   a. Who participates
   b. Their responsibilities
   c. What events are scheduled that they need to take part in

   Project procedures are designed to inform project participants how to participate in the process. They should contain appropriate tool guidance and any information necessary for the participants to execute the process.

   Each document should build on information in the previous document, but all three documents (or at least all of the information in the three documents) should be produced for each project charter.

Creating Project Approaches, Plans, and Procedures

1. Approach (WHAT)
   a. Description:
      i. Importance
      ii. Purpose
      iii. Critical success factors
   b. Overview of approach
2. Plan (WHEN and WHO)
   a. Who participates
   b. Who approves
   c. Responsibilities
   d. Schedule of events
3. Procedures (HOW)
   a. How to participate
   b. How to approve
   c. Toolsets used
   d. Guidelines for use

---

8. Negotiate and agree on the project charter details:
   a. Negotiate the charter details.

      When negotiating, make sure you are listening to the sponsor. The project charter should reflect the desires of the business organization, not the technical desires of the project team.
   b. Estimating is not negotiating.

      Do not use the estimates as a negotiating tool. Estimate the work as you see it, and then work with the project sponsor to determine how the work can be modified (scope reduction, resources added, etc.) to make the charter agreeable. If you rework the estimates to make the charter fit the budget, you will be sending a message to the project sponsor that the numbers are arbitrary and can be changed. This often will lead to projects that have unrealistic estimates and, therefore, are doomed to failure from the start.

      When negotiating, you are working with three aspects of the project: the work (scope), cost (budget), and time (schedule). See the example in Figure 32.2. These aspects relate to each other like the sides of a triangle. You cannot change one side of this triangle without also modifying the other sides. For example, if you

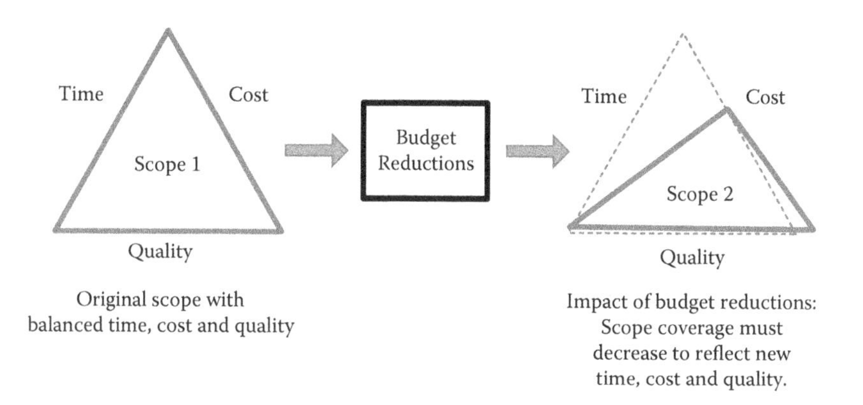

FIGURE 32.2
Estimating is not negotiating.

decrease the budget, you also must decrease the work or increase the time. Work with the project sponsor to devise a project charter that is mutually agreeable.

9. Obtain approval of the project charter.

   Obtain final sign off from the project sponsor on the contents of the project charter.

Software

Some commercial software available includes, but is not limited to:

- Scribd
- Swiftlight®

33

Project Management

DEFINITION

Project management is the application of knowledge, skills, tools, and techniques to project activities in order to meet or exceed stakeholders' needs and expectations from a project. (*PMBOK Guide*, 1996)

Its Uses

This methodology is used to minimize the probability of a project/program not being completed on time, on budget, and producing the desired results.

General

There are two primary and globally accepted Project Management Body of Knowledge and accepted practices. They are the *PMBOK®* (*Project Management Body of Knowledge*) *Guide*, published by the Project Management Institute, and PRINCE2 (**PR**ojects **IN** **C**ontrolled **E**nvironments, ver. 2), which is primarily used in Europe. Because the *PMBOK Guide* is the most widely used, this chapter on project management is based upon the PMBOK.

Preparation

Performance improvement occurs mainly as a result of a number of large and small projects that are undertaken by the organization. These projects involve all levels within the organization and can take less than a hundred hours or millions of hours to complete. They are a critical part of the way an organization's business strategies are implemented. It is extremely important that these multitudes of projects are managed effectively if the stakeholders' needs and expectations are to be met. This is made more complex because conflicting demands are often placed upon the project. For example:

- Scope
- Time
- Cost
- Quality
- Stakeholders with different identified (needs) and unidentified (expectations) requirements

Because there are so many different stakeholders often involved, it places conflicting requirements on a single project. For example, management wants the project to reduce labor cost by 80 percent and organized labor wants it to create more jobs.

The Project Management Institute in Upper Danbury, Pennsylvania, is the leader in defining the body of knowledge for project management. Their PMBOK approach to project management has been widely accepted throughout the world. In addition, the International Organization for Standardization's Technical Committee 176 has released an international standard ISO/DIS 10006: *Guidelines to Quality in Project Management.* These two methodologies complement each other and march hand-in-hand with each other.

A project is a temporary endeavor undertaken to create a unique product or service. A program is a group of related projects managed in a coordinated way. Programs usually include an element of ongoing activities. Large projects are often managed by professional project managers who have no other assignments. However, in most organizations, individuals who serve as project managers are only assigned to spend a small percentage of their time managing many projects. In either case, the individual project manager is responsible for defining a process by which a project is initiated, controlled, and brought to a successful conclusion. This requires the following:

- Project completed on time
- Project completed in budget
- Outputs met specification
- Customers are satisfied
- Team members gain satisfaction as a result of the project

A good project manager follows General George S. Patton's advice when he said, "Don't tell soldiers how to do something. Tell them what to do and you will be amazed at their ingenuity." Although a single project life cycle is very difficult to get everyone to agree to, the project life cycle defined in U.S. DOD's document 5000.2 (Revision 2-26-92 entitled Representative Life Cycle for Defense Acquisition) provides a reasonably good starting point. It is divided into five phases (Figure 33.1).

A life cycle that we like better for an organization that is providing a product and service includes the following phases (Figure 33.2):

- Phase I: Concept and Definition
- Phase II: Design and Development
- Phase III: Creating the Product or Service
- Phase IV: Installation
- Phase V: Operating and Maintenance
- Phase VI: Disposal

Traditionally, projects have followed a pattern of phases from concept to termination. Each phase has particular characteristics that distinguish it from the other phases. Each phase forms part of a logical sequence in which the fundamental and technical specification of the end product or service is progressively defined.

The successful project manager understands that there are four key factors that have to be considered when the project plan is developed. All four factors overlap to a degree, but should first be considered independently and then altogether (Figure 33.3).

In order to effectively manage a project, the individual assigned will be required to address the 10 elements, which are defined in Figure 33.4. The Project Management Institute (PMI) includes only the first nine elements.

Phase 0-Concept Exploration and Definition
Phase I-Demonstration and Validation
Phase II-Engineering and Manufacturing Development
Phase III-Production and Deployment
Phase IV-Operations and Support

FIGURE 33.1
Five phases of project life cycle (PLC). (Representative Life Cycle for Defense Acquisition, per US DOD 5000.2.)

Phase					
I	II	III	IV	V	VI
Concept & Design	Design & Development	Manufacturing	Installation	Operation & Maintenance	Disposal

- New product opportunities
- Analysis of system concept and options
- Product selection
- Technology selection
- Make/buy decisions
- Identify cost drivers
- Construction assessment
- Manufacturability assessment
- Warranty incentives

- Design trade-offs
- Source selection
- Configurations and change controls
- Test strategies
- Repair/throwaway decisions
- Performance tailoring
- Support strategies
- New product introduction

- System integration and verification
- Cost avoidance/cost reduction benefits
- Operating and maintenance cost monitoring
- Product modifications and service enhancements
- Maintenance support resource allocation and optimization

- Retirement cost impact
- Replacement/renewal schemes
- Disposal and salvage value

FIGURE 33.2
Sample applications of project life cycle: Typical activities that are taking place during the individual phases in the project cycle.

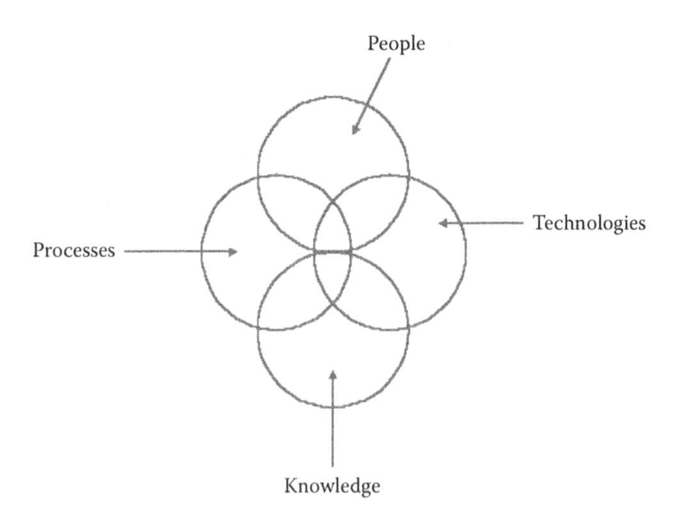

FIGURE 33.3
The key program management factors.

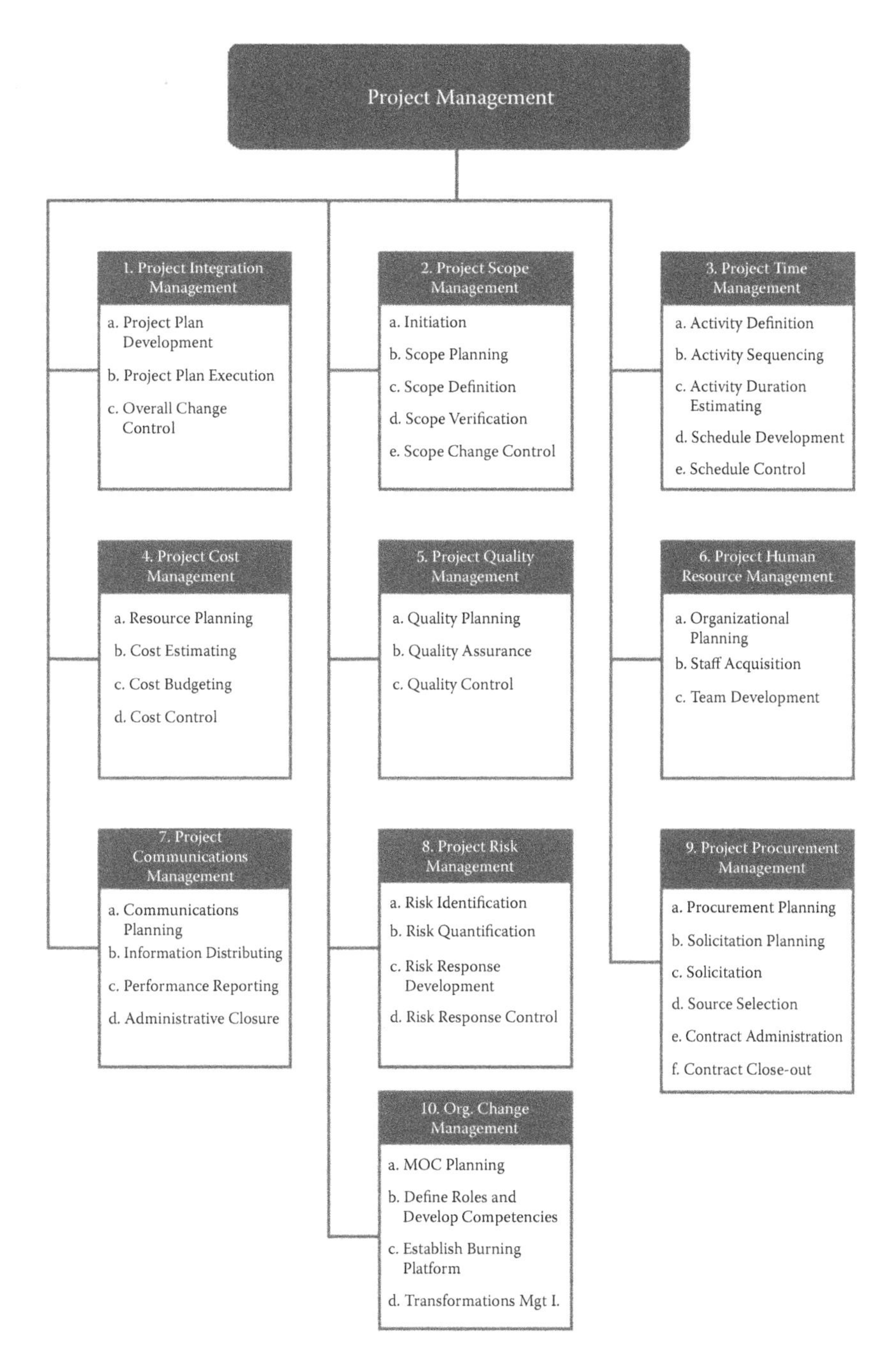

FIGURE 33.4
The 10 elements needed to manage a project.

PMI does not include Organizational Change Management (OCM) as a separate element, but it is placed under the element entitled *Project Risk Management.* Of course, the depth and detail that each element needs in order to be evaluated and managed will vary greatly depending on the scope and complexity of the project.

The book, *A Guide to the Project Management Body of Knowledge*, published by the Project Management Institute (1996) summarizes the project management knowledge areas as follows:

- Project Integration Management

  A subset of project management that includes the processes required to ensure that the various elements of the project are properly coordinated. It consists of:
  - Project plan development: Taking the results of other planning processes and putting them into a consistent, coherent document.
  - Project plan execution: Carrying out the project plan by performing the activities included therein.
  - Overall change control: Coordinating changes across the entire project.
- Project Scope Management

  A subset of project management that includes the processes required to ensure that the project includes all the work required, and only the work required, to complete the project successfully. It consists of:
  - Initiation: Committing the organization to begin the next phase of the project.
  - Scope planning: Developing a written scope statement as the basis for future project decisions.
  - Scope definition: Subdividing the major project deliverables into smaller, more manageable components.
  - Scope verification: Formalizing acceptance of the project scope.
  - Scope change control: Controlling changes to project scope.
- Project Time Management

  A subset of project management that includes the processes required to ensure timely completion of the project. It consists of:
  - Activity definition: Identifying the specific activities that must be performed to produce the various project deliverables.
  - Activity sequencing: Identifying and documenting interactivity dependencies.

    - Activity duration estimating: Estimating the number of work periods that will be needed to complete individual activities.
    - Schedule development: Analyzing activity sequences, activity durations, and resource requirements to create the project schedule.
    - Schedule control: Controlling changes to the project schedule.
- Project Cost Management

  A subset of project management that includes the processes required to ensure that the project is completed within the approved budget. It consists of:
    - Resource planning: Determining what resources (people, equipment, materials) and what quantities of each should be used to perform project activities.
    - Cost estimating: Developing an approximation (estimate) of the costs of the resources needed to complete project activities.
    - Cost budgeting: Allocating the overall cost estimate to individual work items.
    - Cost control: Controlling changes to the project budget.
- Project Quality Management

  A subset of project management that includes the processes required to ensure that the project will satisfy the needs for which it was undertaken. It consists of:
    - Quality planning: Identifying which quality standards are relevant to the project and determining how to satisfy them.
    - Quality assurance: Evaluating overall project performance on a regular basis to provide confidence that the project will satisfy the relevant quality standards.
    - Quality control: Monitoring specific project results to determine if they comply with relevant quality standards and identifying ways to eliminate causes of unsatisfactory performance.
- Project Human Resource Management

  A subset of project management that includes the processes required to make the most effective use of the people involved with the project. It consists of:
    - Organizational planning: Identifying, documenting, and assigning project roles, responsibilities, and reporting relationships.
    - Staff acquisition: Getting the human resources needed assigned to and working on the project.
    - Team development: Developing individual and group skills to enhance project performance.

- Project Communications Management

  A subset of project management that includes the processes required to ensure timely and appropriate generation, collection, dissemination, storage, and ultimate disposition of project information. It consists of:
  - Communications planning: Determining the information and communications needs of the stakeholders: who needs what information, when will they need it, and how will it be given to them.
  - Information distribution: Making needed information available to project stakeholders in a timely manner.
  - Performance reporting: Collecting and disseminating performance information. This includes status reporting, progress measurement, and forecasting.
  - Administrative closure: Generating, gathering, and disseminating information to formalize phase or project completion.
- Project Risk Management

  A subset of project management that includes the processes concerned with identifying, analyzing, and responding to project risk. It consists of:
  - Risk identification: Determining which risks are likely to affect the project and documenting the characteristics of each.
  - Risk quantification: Evaluating risks and risk interactions to assess the range of possible project outcomes.
  - Risk response development: Defining enhancement steps for opportunities and responses to threats.
  - Risk response control: Responding to changes in risk over the course of the project.
- Project Procurement Management

  A subset of project management that includes the processes required to acquire goods and services from outside the performing organization. It consists of:
  - Procurement planning: Determining what to procure and when.
  - Solicitation planning: Documenting product requirements and identifying potential sources.
  - Solicitation: Obtaining quotations, bids, offers, or proposals as appropriate.
  - Source selection: Choosing from among potential sellers.
  - Contract administration: Managing the relationship with the seller.
  - Contract closeout: Completion and settlement of the contract, including resolution of any open items.

- Organizational Change Management (OCM)

  This is a part of project management that is directed at the people side of the project. OCM helps prepare the people who either live in the process that is being changed or have their work lives changed as a result of the project not to resist the change.

  OCM often prepared the employees so well that they look forward to the change. (Note: This is not part of the PMBOK project management concept).
  - OCM planning: Define the level of resistance to change and prepare a plan to offset the resistance.
  - Define roles and develop competencies: Identify who will serve as sponsors, change agents, change targets, and change advocates. Then, train each individual on how to perform the specific role.
  - Establish burning platform: Define why the as-is process needs to be changed and prepare a vision that defines how the as-is pain will be lessened by the future-state solution.
  - Transformations management: Implement the OCM plan. Test for black holes and lack of acceptance. Train affected personnel in new skills required by the change.

Examples

Table 33.1 describes the approaches required to do just one part of managing a project risk management. With a list of approaches so long, it is easy to see that managing a project is not for the weak of heart or the inexperienced.

Project Plan Example

Title Page

1. Foreword
2. Contents, distribution, and amendments record
3. Introduction
   - 3.1. General description
   - 3.2. Scope
   - 3.3. Project equipment
   - 3.4. Project security and privacy
4. Project aims and objectives
5. Project policy
6. Project approvals required and authorization limits

7. Project organization
8. Project harmonization
9. Project implementation strategy
   9.1. Project management philosophy
   9.2. Implementation plans
   9.3. System integration
   9.4. Completed project work
10. Acceptance procedure
11. Program management
12. Procurement strategy
13. Contract management
14. Communications management
15. Configuration management
    15.1. Configuration control requirements
    15.2. Configuration management system
16. Financial management
17. Risk management
18. Project resource management
19. Technical management
20. Test and evaluation
21. Reliability management
    21.1. Availability, reliability, and maintainability (ARM)
    21.2. Quality management
22. Quality plan (A typical work breakdown structure in support of the project plan is shown in Table 33.2.)

TABLE 33.1

Methods Used in Risk Analysis

Method	Description and Usage
Event Tree Analysis	A hazard identification and frequency analysis technique that employs inductive reasoning to translate different initiating events into possible outcomes.
Fault Mode & Effects Analysis & Fault Mode Effect & Criticality Analysis	A fundamental hazard identification and frequency analysis technique that analyzes all the fault modes of a given equipment item for their effects both on other components and the system.
Fault Tree Analysis	A hazard identification and frequency analysis technique that starts with the undesired event and determines all the ways in which it could occur. These are displayed graphically.

TABLE 33.1 ***(Continued)***

Methods Used in Risk Analysis

Method	Description and Usage
Hazard & Operability Study	A fundamental hazard identification technique that systematically evaluates each part of the system to see how deviations from the design intent can occur and whether they can cause problems.
Human Reliability Analysis	A frequency analysis technique that deals with the impact of people on system performance and evaluates the influence of human errors on reliability.
Preliminary Hazard Analysis	A hazard identification and frequency analysis technique that can be used early in the design stage to identify hazards and assess their criticality.
Reliability Block Diagram	A frequency analysis technique that creates a model of the system and its redundancies to evaluate the overall system reliability.
Category Rating	A means of rating risks by the categories in which they fall in order to create prioritized groups of risks.
Checklists	A hazard identification technique that provides a listing of typical hazardous substances and/or potential accident sources, which need to be considered. Can evaluate conformance with codes and standards.
Common Mode Failure Analysis	A method for assessing whether the coincidental failure of a number of different parts or components within a system is possible and its likely overall effect.
Consequence Models	The estimation of the impact of an event on people, property, or the environment. Both simplified analytical approaches and complex computer models are available.
Delphi Technique	A means of combining expert opinions that may support frequency analysis, consequence modeling, and/or risk estimation.
Hazard Indices	A hazard identification/evaluation technique that can be used to rank different system options and identify the less hazardous options.
Monte Carlo Simulation and other simulation techniques	A frequency analysis technique that uses a model of the system to evaluate variations in input conditions and assumptions.
Paired Comparisons	A means of estimation and ranking a set of risks by looking at pairs of risks and evaluating just one pair at a time.
Review of Historical Data	A hazard identification technique that can be used to identify potential problem areas and also provide an input into frequency analysis based on accident and reliability data.
Sneak Analysis	A method of identifying latent paths that could cause the occurrence of unforeseen events.

TABLE 33.2

Project Process as it Relates to the Phases in a Project

	Phases			
Processes	**Conception**	**Development**	**Realization**	**Termination**
Strategic project process				
Strategic project process	A	x	x	x
Operational process groups and processes within groups				
Scope-related operational processes				
Concept development	A			
Scope definition	A	x		
Task definition	x	A	x	
Task realization		A	A	x
Change management		A	A	
Time-related operational processes				
Key event schedule planning	x	A	x	
Activity dependency planning	x	A		
Duration estimation	x	A		
Schedule development		A	x	
Schedule control		x	A	x
Cost-related operational processes				
Cost estimation	A	x		
Budgeting		A	x	
Cost control		x	A	x
Resource-related operational processes (except personnel)				
Resource planning	x	A		x
Resource control		x	A	x
Personnel-related operational processes				
Organizational structure definition	x	A	A	
Responsibility identification and assignment	x	A	x	
Staff planning and control		x	A	x
Team building	x	A	A	x

TABLE 33.2 *(Continued)*

Project Process as it Relates to the Phases in a Project

	Phases			
Processes	**Conception**	**Development**	**Realization**	**Termination**
Communication-related operational processes				
Communication planning	x	A		
Meeting management	x	A	A	x
Information distribution		A	A	x
Communication closure			x	A
Risk-related operational processes				
Risk identification	A	A	x	
Risk assessment	A	A	x	
Solution development		A	x	
Risk control		x	A	
Procurement-related operational processes				
Procurement planning	x	A		
Requirements documentation	x	A		
Supplier evaluation	x	A		
Contracting		A	x	
Contract administration		x	A	x
Product-related operational processes				
Design	x	A		
Procurement	x	A		
Realization			A	
Commissioning				A
Integration-related operational processes				
Project plan development	A			
Project plan execution		A	A	A
Change control		A	A	
Supporting processes				

Note: A = Key process in the phase; x = Applicable process in the phase.

Software

There are many excellent project management software packages. The following is a list of just a few of them (Figure 33.5 is a block diagram of a typical project management process).

- Replicon®
- Hydra™
- NetSuite® Open Air

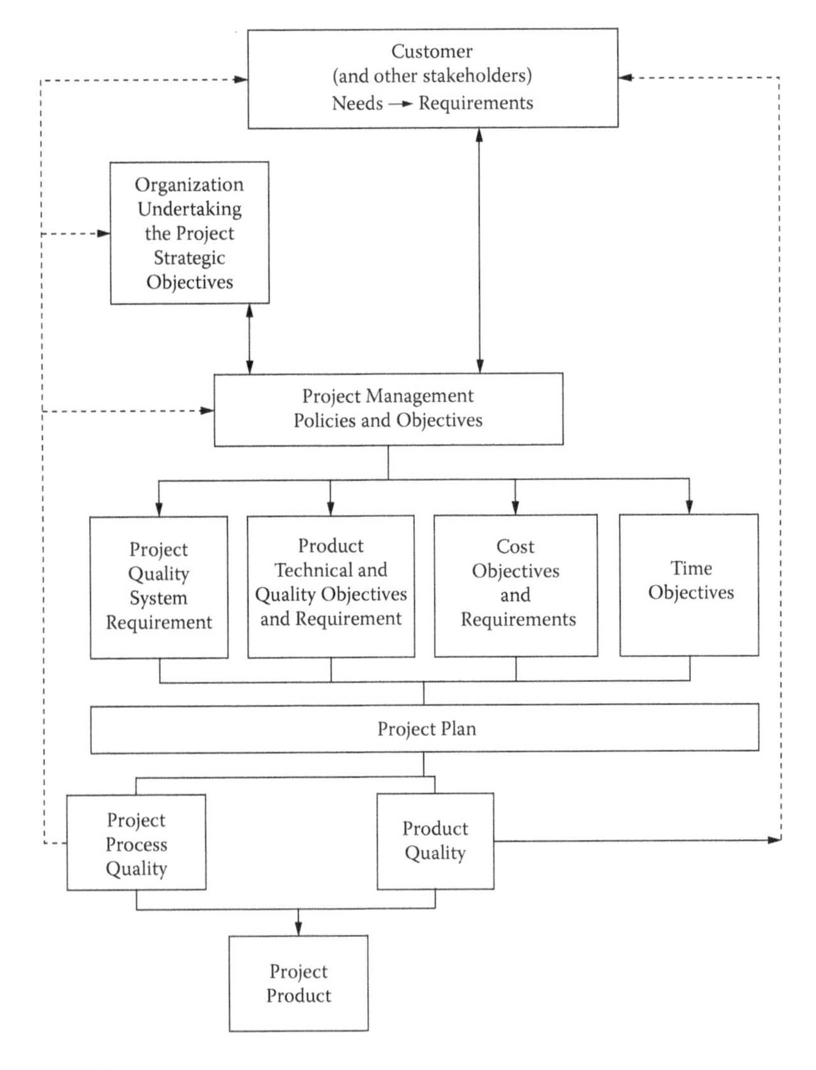

FIGURE 33.5
Block diagram of the project management process.

- Clearview™
- Gazelle®
- Vorex®
- VPO™
- Deltek Project Doc Control®
- Clarizen®

ADDITIONAL READING

Badiru, A. B., and G. E. Whitehouse. 1989. *Computer tools, models, & techniques for project management.* Blue Ridge Summit, PA: TAB Books.

Block, R. 1983. *The politics of projects.* New York: Yourdon Press.

Brooks, H. E., Jr. 1989. *Project management in the information technology age.* Downey, CA: Sterling Series.

Cleland, D., and R. Gareis, eds. 1994. *Global project management handbook.* New York: McGraw-Hill.

Dinsmore, P. 1990. *Human factors in project management.* New York: AMACOM Publications.

Dinsmore, P. C. 1993. *The AMA handbook of project management.* New York: AMACOM Books.

Einsiedel, A. 1988. *Improving project management.* Englewood Cliffs, NJ: Prentice-Hall.

Focardi, S., and C. Jonas. 1998. *Risk management: Framework, methods and practice.* New York: McGraw-Hill.

International Organization for Standardization. 1998. *ISO/DIS 10006: Guidelines to quality in project management.* Geneva: IOS.

PMI Standards Committee. 1996. *A guide to the project management body of knowledge.* Upper Darby, PA: Project Management Institute.

Project Management Institute. 1996. *A guide to the project management body of knowledge.* Sylva, NC: Project Management Institute.

Westney, R. E. 1992. *Computerized management of multiple small projects.* New York: Marcel Dekker, Inc.

34

Risk Management

DEFINITION

Risk management is the identification, assessment, and prioritization of risks (defined in ISO 31000 as the effect of uncertainty on objectives, whether positive or negative), followed by coordinated and economical application of resources to minimize, monitor, and control the probability and/or impact of unfortunate events.

Its Uses

Risk management is used in the planning phase of a process, product, or service as well as after problems occur to identify possible situations that may cause problems with the process, product, or service.

Risk management can be applied to many areas including management, security, engineering, industrial processes, financial portfolios, actuarial assessments, public health and safety, and service areas.

General

There are several actions to consider in managing risk, which include:

- Contingent Action: A preplanned response designed to minimize or eliminate the impact of a known risk factor on the project should it occur. Contingent actions are reactive responses to risk factors.
- Risk Assessment: A process that assigns a probability of occurrence and severity of consequence to a risk.
- Project Risk: A characteristic of a specific project, a circumstance of the project, or a feature of its environment that is recognized to have a potentially adverse effect on the project or the quality of

its deliverables. A risk factor can be viewed from two dimensions: risk probability and risk severity. Risk probability is the estimated likelihood that a certain risk will occur during the project. If the risk occurs, risk severity is an estimate of the extent of its negative impact on the project.

- Risk Factor: A characteristic of a project that is recognized to have a potentially adverse effect on the project or the quality of its deliverables.
- Risk Management Strategy: A means of minimizing the likelihood and impact of a known risk factor on the project. Risk management strategies are preventive and proactive actions by the project manager.

Dealing with Project Risk

Risk is inevitable on any project. All projects have some degree of uncertainty in their assumptions and in the environment in which they are executed. Project risks cannot be eliminated entirely, but a large proportion of them can be anticipated and reduced. Risk management systematically identifies the specific risk factors for a project and then establishes an explicit risk management strategy to reduce the known risks of the project. The risk management strategy defines safeguards to minimize the probability that certain risks will materialize and contingent actions to deal with the risks if they do occur.

Assessing Project Risk

Project risk assessment is best accomplished as a group effort in order to fully explore the potential risk of the project. At a minimum, the project manager and project sponsor should participate, but useful insights can be gained by involving other participants who do not have a direct stake in the project.

Preparation

- Start by identifying all the ways that a project, process, product, or service can fail or have a problem.
- Next, evaluate the probability of each of these occurring.

- Then, determine your ability to detect each of the scenarios.
- Once this is done, each possible failure mode can be ranked for its effect on the process, product, or service.
- Based on this ranking, a mitigation or elimination plan can be put in place together with a process for continually evaluating the effectiveness of the process, product, or service.

The Failure Mode and Effects Analysis (FMEA) is an effective tool for evaluating risk and development after a problem has occurred.

When to Use FMEA

- When a process, product, or service is being designed or redesigned
- When identifying the required control plans for a new or modified process
- When identifying improvement goals for an existing process or product
- When analyzing failures of an existing process or product
- Repeated throughout the life of the process or product

FMEA Procedure

The FMEA procedure is general and can be applied to any process or product to identify possible failure modes and establish activities to mitigate or eliminate them.

The steps in creating an FMEA include:

1. Use a cross-functional team with diverse knowledge about the process or product.
2. Identify the scope of the process or product to be evaluated for the FMEA.
3. Fill in the identifying information at the top of your FMEA form.
4. Identify what will be included in your scope. Ask:
    a. What is the purpose of this system, design, process, or service?
    b. What do our customers expect it to do?
5. Identify all the ways failure could happen for each process step.
6. For each failure mode, identify all the consequences, minor problem to total failure.

7. Assign a number from 1 to 10 to identify the seriousness of the consequence. **(S)**
8. For each failure mode, determine all the potential root causes.
9. For each cause, determine the probability of occurrence and assign a number from 1 to 10 to identify how often you expect to observe the problem. **(O)**
10. Establish the ability of current controls on the process to detect the occurrence and mitigate or eliminate it. Assign a value from 1 to 10 to indicate the ability to detect an occurrence, with 10 being "can't detect" and 1 being "easily detected." **(D)**
11. Calculate the risk priority number, or RPN, and criticality numbers:
    a. RPN = Seriousness (S) × Occurrence (O) × ability to Detect (D).
    b. Criticality = (S) × (O). These numbers provide guidance for ranking potential failures in the order they should be addressed.
12. For each high RPN number, usually over 300, identify recommended actions.
13. Based on the actions taken, observe the results and recalculate a new RPN number.

Example

See Table 34.1 and Table 34.2.

In this example there are three areas that have to be addressed: reservation wait time, insufficient tables, and poor food.

Software

Some commercial software available includes, but is not limited to:

- ProcessGene GRC Software Suite™
- Procurify™
- Apttus Contract
- QI Macros®
- Minitab®
- JMP®
- Sigma XL®

TABLE 34.1

FMEA Process Form

Item/ Process	Cafe	Process Responsibility:	FMEA Number	
Subsystem		Key Date:	Page	1 of 1
Model Years			Prepared by:	
Core Team:	Who		FMEA Date	

Item Function	Potential Failure Mode	Potential Effect(s) of Failure	Severity	Potential Cause(s)/ Mechanism(s) of Failure	Occur	Current Process Controls Prevention	Current Process Controls Detection	Detect	R. P. N.
Reservation	Too long of a wait	Customer leaves	8	Not enough staff	5	Four people on front desk	None	8	320
	No tables empty	Customer leaves	8	Insufficient tables 6	Schedule by reservation	Reservations	6	384	
Poor Food	Substandard food	Spoiled or out of date	10	Too long held	6	Rotate food	None	8	480
Service	Not served in 20 minutes	Customers anxious	6	Slow order processing	5	Number of chefs	Schedule chefs	5	150

TABLE 34.2

FMEA Scoring Guide

Severity of Effect	Occurrence Rating	Detection
1. None	1. Remote <.01/1000	1. Almost Certain
2. Very Minor	2. Low – 0.1/1000	2. Very High
3. Minor	3. Low – 0.5/1000	3. High
4. Very Low	4. Moderate – 1/1000	4. Moderate High
5. Low	5. Moderate – 2/1000	5. Moderate
6. Moderate	6. Moderate – 5/1000	6. Low
7. High	7. High – 10/1000	7. Very Low
8. Very High	8. High 20/1000	8. Remote
9. Hazardous with warning	9. Very High 50/1000	9. Very Remote
10. Hazardous without warning	10. Very High >100/1000	10. Absolute Uncertainty

35

Role Mapping

DEFINITION

Role mapping is a method of graphically defining the relationship of people relative to their professional relationships, political, and organizational structures that are necessary to the success of the various components of a major change.

Its Uses

It is used to help identify the critical personnel and their relationship to the success or failure of a change in organization, technology, and political elements inherent in any change.

General

Role mapping is a means of defining the infrastructure to be used in accomplishing a change. The elements involved include:

- Identify the people involved in securing the necessary sponsorship for a change to succeed
- Identify the constituencies affected by the change
- Generate a visual picture of the individuals, groups, and interrelationships that must be orchestrated to accomplish the change
- Understand the "flow" of how the change will unfold within the organization
- Build a common understanding of the issues, problems, and opportunities inherent in accomplishing the change

- Represent the political terrain of the change environment that must be addressed when developing the implementation plans
- Better understand the dynamics of influence that may affect the outcome of the change effort

Role mapping is best developed in group discussions among the parties responsible for analyzing, planning, and implementing a change. You should consider using a facilitated session to build the role map.

Preparation

A role map not only needs to depict what roles (advocate, sponsor, agent, or target) each person currently possesses, but also the roles each one needs to fulfill in order for the change to be successful. Once these two states are established (the present and the desired), it is necessary to develop strategies to transform people into their needed roles. Until actions are taken and these necessary roles are established, the change cannot be completed successfully.

Conducting Role Mapping

- Conduct a facilitated session.

  Conduct a facilitated session with the key decision makers of the change (i.e., executive sponsor team).
- Identify the components required for a successful change project.

  Identifying components may be by identifying different constituency groups that reflect different terrains of influences or by breaking your project into phases, identifying each phase as a component. List the key components.

  The roles involved include the following:
  - Change Advocate: Individual or group who wants to achieve a change but who lacks sufficient sponsorship.
  - Change Agent: An individual or group who is responsible for implementing change.
  - Change (or sometimes called *change target*): An individual or group who must actually change.
  - Executive Sponsor: The person who has ultimate authority over and responsibility for the project. The executive sponsor

has a vested interest in the results of the project, he or she is the person who:
  - funds the project;
  - resolves conflicts over policy or objectives; and
  - provides high-level direction.

  The executive sponsor also is responsible for approving changes to the system during the development process, and for providing whatever additional funds those changes require. Often the executive sponsor delegates day-to-day participation to the project sponsor. Occasionally, the executive sponsor and the project sponsor will be the same individual.
- Initiating Sponsor: Individual or group who has the power to initiate and legitimize the change for all the affected targets.
- Sponsor: An individual or group with the authority to legitimize change and/or who has the political, logistical, and economic proximity to the change targets.
- Sustaining Sponsor: A person or group who has direct influence with the targets, based on the political, logistical, and economic proximity to the targets.

Start by identifying the initiating sponsor. This is the person who can sign for the resources that will be required to successfully accomplish the project. This individual is usually a high-level manager. If it's a team that needs to approve the project to get it funded and staffed, then the initiating sponsor would be the chairman of the team.

The next step is to identify the highest level manager that meets the requirements to be a sustaining sponsor. If that individual is not immediately identifiable, you should have a discussion with the initiating sponsor and have him or her identify the highest level manager that she or he is holding responsible for ensuring that the project is implemented successfully. This manager is initially a changee who needs to be converted from a target to a sustaining sponsor. This process is repeated over and over again until the individual who was transformed from a target into a sustaining sponsor has managers reporting to him/her, and these individuals are truly the targets that will be impacted by the change initiative. In that manner, you basically established the skeleton that the role map is built upon as you have identified the initiating sponsor, the sustaining sponsor, and targets.

Identifying the advocates presents a different challenge. Now you have to look at the sustaining sponsors and identify individuals who have the ability to influence the sustaining sponsor and who are very interested in seeing the project successfully completed. These are individuals that are potential advocates. Change agents are individuals who are assigned to the project that have been trained in using the organizational change management methodologies tools and techniques.

You can see establishing a role map requires good understanding of the organizational structure and how the project will impact the organization. See Figure 35.1 for an example of a role map.

Example

- Map each component separately once they are determined.

  A role map diagram is drawn for each component. The total role map diagram of an entire change project is formed by aggregating the individual component diagrams. Summarize the key people and relationships among the various maps.
- Confirm the role map with the clients.

Additional guidelines for role mapping:

- Role maps should be built throughout the change project, although it is imperative that one is completed early in the life of the project. It helps clarify the true nature and size of the work effort, and it identifies the key constant constituents that require special care when announcing the change.
- Role mapping is a means of defining the infrastructure that will affect a change project.
- A role map not only needs to depict what roles (advocate, sponsor, agent, or target) each person currently possesses, but also the roles each one needs to fulfill in order for the change to be successful. Once these two are established (the present and the desired), one must analyze the current structure of the organization to develop strategies to transform these people into their needed roles. Until actions are taken in these necessary roles, which are established, implementation of the change cannot successfully occur.
- Various role maps may be created throughout the life of the project. The ultimate determination of how the map is generated revolves around what stage of the project you are in and what your goal is

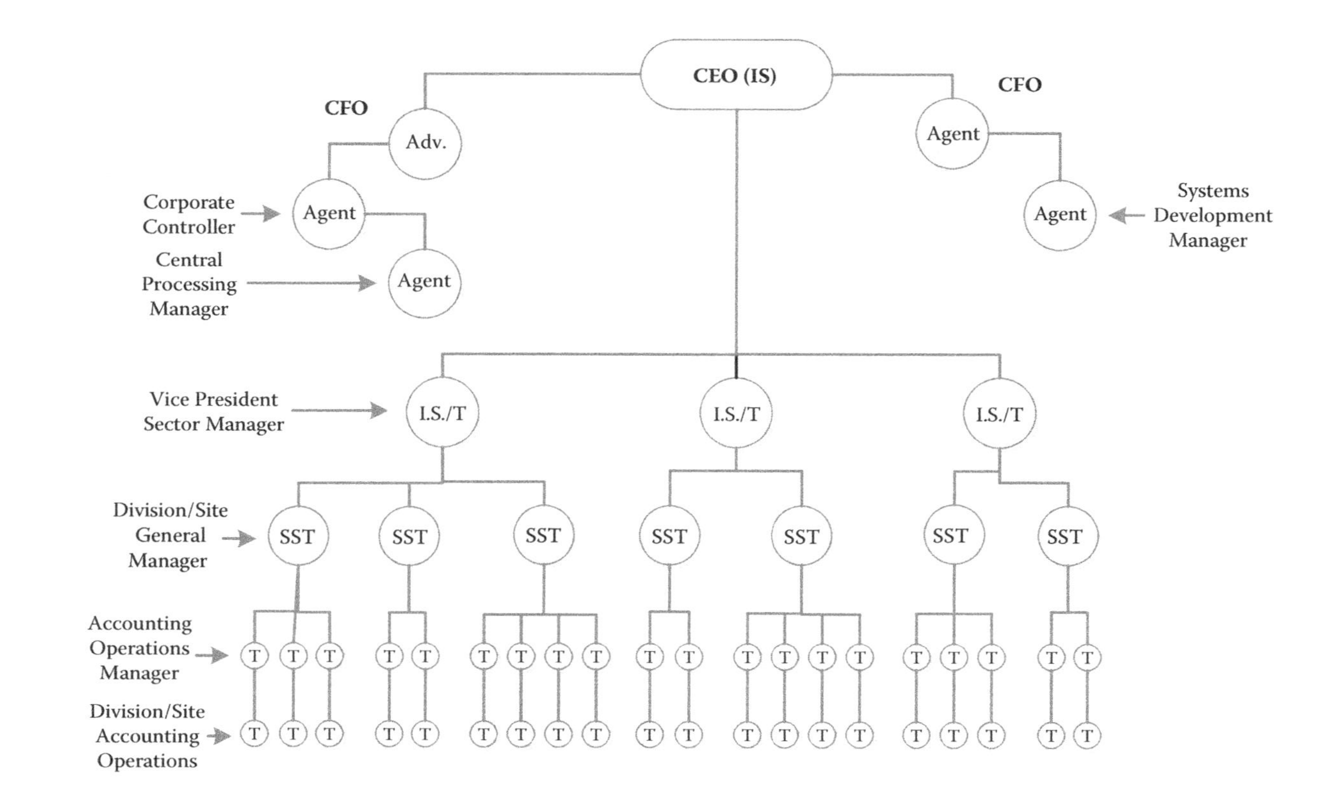

FIGURE 35.1
A typical role map.

at that stage. Who do you need in what role and what do you want from them? How do you go about it (the chain of command)?

- Anyone can generate a map of the change project's key roles at any time in the life of that project. Nevertheless, it is imperative that a role map diagram be developed early in the project. This helps clarify the true nature and size of the effort. Role maps should be generated within the project so that everyone clearly understands their responsibilities. Role mapping is also an effective tool to validate that the communication system is working. It is extremely important that all of the sustaining sponsors be actively involved in the project and that they are projecting a continuous positive view of the project.
- The role map diagram should reflect the true terrain of influences related to a given condition. Do not assume that the formal organizational chart always depicts this. For example, in some cases, it is the informal or unofficial relationships that are the most crucial to the successful implementation of change. Other times, it can be the relationships outside the organization that are important influential factors (e.g., unions, associations).
- To help avoid the tendency of simply duplicating a formal organizational hierarchy, always start the map at the bottom. The first question to ask is: "Who are the ultimate targets of the change?" Ask next: "To whom will these targets look for legitimization or sanctioning messages (sustaining sponsors)?" Finally ask: "Who will sponsor these people?" Role map diagrams are always developed from the bottom up.
- Avoid developing a map only around strong or weak players in the various roles. For example, don't exclude a sustaining sponsor simply because you take for granted his or her support for the change. Include all players who are involved in key roles regardless of their current commitment or predisposition to the change.
- Be careful. It is not always easy to tell who is the initiating sponsor (IS). The IS may not be the person who originates the idea; that person is the initiating advocate. The IS is the person or group who has the organizational power to break from the status quo and officially sanction the change. One way to tell the difference between the sustaining sponsor (SS) and the (IS) is that the SS cannot start a change without first gaining permission from his or her sponsor. This means that the person must first function as an advocate to his/her boss. Initiating sponsors do not ask for permission to engage change;

instead, they keep their boss informed of what they are going to do or what they have already done regarding a change.

- Avoid the tendency to view powerful advocates as sponsors. Advocates are individuals who want to achieve a change or who are successful in convincing others of its necessity. Due to their position in the organizational structure, they do not have people reporting directly to them that will be significantly impacted by the implementation of the change and, as a result, they cannot be considered as initiating or sustaining sponsors.
- Look for recurring relationship patterns among the various component role map diagrams of an overall change project. In some cases, the entire map developed for one component may be applied without any modification to another component. These are referred to as clones.
- Role maps should be updated or evaluated throughout the life of the project to identify new players and evaluate effectiveness. The role maps should reflect the current organization, but as the project progresses, they may reflect the future.

Software

Some commercial software available includes, but is not limited to:

- SmpMapX 20

36

Root Cause Analysis

DEFINITION

Root Cause Analysis is the process of identifying the various causes affecting a particular problem, process, or issue, and determining the real reasons that caused the condition.

Its Uses

It provides the basis for eliminating or streamlining activities or subprocesses.

General

A Root Cause Analysis identifies the cause of a given condition or outcome, thereby providing the necessary information to mitigate and eliminate the cause and, therefore, the condition or outcome.

Preparation

- Identify performance issues to be analyzed.

  Identify issues in processes, such as process performance, information processing, and infrastructure.

  Assign teams of process owners, executive managers, or both. Where appropriate, make cross-functional assignments. Teams should be responsible for the analysis of one or more activities within the subprocess boundaries.
- Classifying root causes.

  Root Cause Analysis seeks, for each activity, to answer the question: Why is this done? Information may be elicited by brainstorming answers to this question. Another effective technique to determine the actual root cause of an activity is to ask: Why is this done? five times to determine an activity's root cause.

- Root causes may be classified as follows:
  - Required Conversion: A required conversion root cause triggers an activity that is required to modify inputs in a manner that generates an output that better meets customer-specified requirements. This is the only type of root cause that can result in a purely value-added activity.
  - Policy/Control: Root causes of this nature occur due to either internal or external organizational regulations. Examples of internal regulations include the use of organization-specific publication formats and logos. Examples of external regulations include government reporting requirements.
  - Inspection: This root cause triggers activities that involve the verification and validation of activity input and output specifications.
  - Rework: Activities that are caused by rework requirements occur when an activity has to be repeated due to substandard initial activity performance.
  - Internal: Internal root causes result in activities that are performed due to insufficient performance or inefficiencies within the organization. Examples include lack of information, insufficient knowledge/training, inconsistent procedures, and data entry errors.
  - External: Activities may be caused by factors external to the organization; these may include causes such as untimely or inaccurate inputs.
  - Systems: Some activities occur because of inadequate systems configurations; root causes of this type include a lack of information or poor system design specifications, functionality, or integration.

    Note: Some of these classifications overlap; they do not represent a rigorous scheme. They are provided to stimulate the root cause identification process. A Root Cause Analysis may result in varying numbers and types of root cause categories.
  - Determine the root causes of the activities or subprocesses.

    Have the teams use facilitation techniques to determine the root causes for their assigned activities. Root causes are identified by investigating the fundamental reason for the execution of the activity. In this process, the primary question they should address is: Why is this done? Document the root causes to support reviews that will identify candidate activities or subprocesses for elimination or streamlining.

Example

It takes seven days to process a loan request. Evaluation of the process indicates that the paperwork passes through eight different evaluation steps. Each step reviews different aspects of the loan application from the original information provided.

Analysis of the process indicated that information from the original paperwork was entered several times into the bank's database.

When an error is found, the paperwork is returned to the applicant for correction.

The analysis indicated that 35 percent of the errors were in entering data into the database.

It was also determined that four of the review operations were redundant.

There were three root causes for the seven-day turnaround time.

1. Multiple data entry points
2. Wait time between evaluations
3. Redundant operations

Addressing these, the following actions were taken:

1. Data entry is done at the beginning of the process.
2. Processing was converted to electronic from paper-based.
3. Three review operations were eliminated.

From these actions, the processing time was reduced from seven days to three days.

Software

Some commercial software available includes, but is not limited to:

- Apollo Root Cause™
- REASON Root Cause®
- TapRooT Cause®

37

Run Charts

DEFINITION

Run charts are a graphic display of data used to assess the stability of a process over time, or over a sequence of events (such as the number of batches produced). A statistical process control (SPC) chart is a type of a run chart. The run chart is the simplest form of a control chart.

Its Uses

They form a simple tool used to determine if there are changes in a process over time. Such processes can be identified as candidates for process improvement.

General

Run charts are an initial way of looking at the performance of a process over a short period of time or over changes to determine its stability or sensitivity to change.

Preparation

Using a Run Chart

- Determine the data capturing process.
- Determine what data are to be captured.
- Decide on the sample size and the frequency of measurement. Also decide on the time horizon over which data are to be captured.

- Design the chart.

  Create the run chart as a modified form of an *X-Y* graph. Label the *X*-axis of the run chart to be the time over which data are collected. Label the *Y*-axis for the data variable.
- Collect and plot the data.
- Start collecting the data. After at least 10 data points have been collected, calculate the median. Plot the median value half way up the *Y*-axis. Draw a centerline across the run chart, parallel to the *X*-axis. Label the line as the median.
- Plot the median value half way up the *Y*-axis. Draw a centerline across the run chart, parallel to the *X*-axis. Label the line as the median.
- Plot each data point as it is collected.
- Analyze the data.

  Analyze the run chart on an ongoing basis as more data points are added. Look for abnormal occurrences in the data plot to determine possible changes in the process.

  Abnormalities can include several consecutive data points falling on the same side of the median, several consecutive data points increasing or decreasing, or many consecutive points alternating between being above the line and being below the line.

Example

See Figure 37.1.

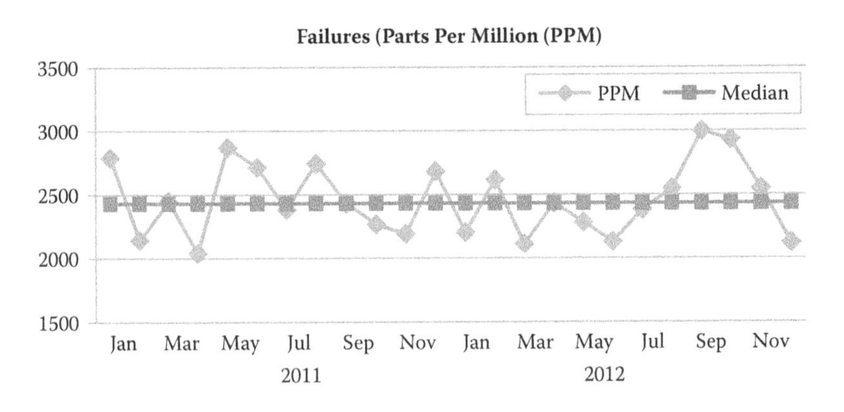

FIGURE 37.1
Sample run chart.

Software

Some commercial software available includes, but is not limited to:

- Minitab®
- JMP®
- QI Macros™
- Sigma XL®

38

Scatter Diagram

DEFINITION

A scatter diagram is a graphic tool used to study the relationship between two variables. Scatter diagrams are used to test for possible cause-and-effect relationships. It does not prove that one variable causes the other, but it does show whether a relationship exists and reveals the character of that relationship.

Its Uses

It is used to test for possible cause-and-effect relationships. It does not prove that one variable causes the other, but it does show whether a relationship exists between variables and the character of that relationship.

General

The diagram is formed by plotting the two variables against each other. The resulting cluster of data points can be analyzed to identify possible relationships. The direction and tightness of the cluster gives an indication as to the type of relationship that exists.

Preparation

- Prepare to examine two variables:
  - Identify two variables to be compared to identify potential relationships. Both variables must have measurable characteristics.
  - Collect paired samples of data on the two variables.
- Create the diagram:
  - Construct the horizontal and vertical axes of the diagram. The vertical axis is usually used for the variable on which you

are measuring the possible effect (the dependent variable). The horizontal axis is used for the variable that is being investigated as the possible cause of the other variable (independent variable).

- Analyze the data to see whether a relationship exists:
  - Plot the data points on the diagram.
- Draw an upper and lower boundary line along the outer region of the data points. Draw a center line between these two boundary lines. This center line (regression line) will give a rough average of the data point relationships.
  Note: If the data points seem to cluster in a curved pattern, do not try to draw boundary lines. Instead, trace in a center line to indicate the trend.
- Analyze the cluster patterns of the data points to determine if a relationship exists, and if so, to determine what type of relationship exists.

Example

See Figure 38.1.

When a scatter diagram shows a strong relationship between two variables, it is tempting to assume that one variable causes the other. Avoid this and similar errors when studying the relationships between two variables. Keep in mind that if a relationship does exist, it is probably only one of many factors that affect the dependent variable. Do not make the generalization that Variable A causes Variable B. Also remember that relationships do not have to be linear or 1 to 1.

Software

Some commercial software available includes, but is not limited to:

- Excel®
- Minitab®
- JMP®
- QI Macros™
- Sigma XL®

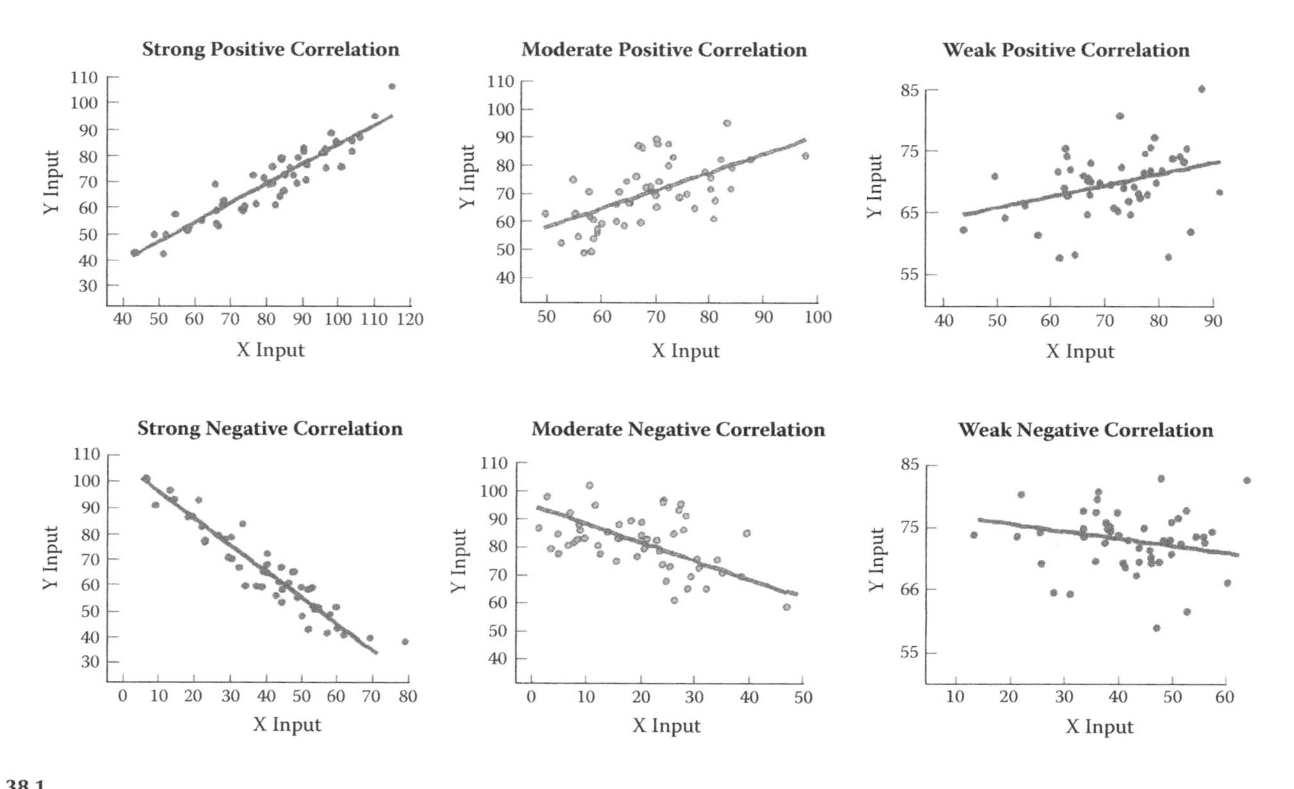

FIGURE 38.1
Sample scatter diagrams.

39

Solutions Evaluation

DEFINITION

Solutions evaluation is a technique used to help review and narrow solutions on the basis of a thorough cost/benefit analysis.

Its Uses

It is designed to help identify all qualitative and quantitative costs and benefits that could potentially be associated with the implementation of each solution.

General

This technique results in an added set of criteria for deciding which solutions should be further considered for implementation. In some cases, conducting the solutions evaluation technique provides a clear choice for implementation; if not, it may be necessary to apply a narrowing technique, such as consensus building or voting, in order to proceed to implementation.

Preparation

- Create a list of solutions to evaluate.

  Using brainstorming or another expanding technique, establish the list of solutions to be considered for implementation.
- Identify customer-related consequences of each solution.

  Conduct a team brainstorming session to identify potential customer-related qualitative impacts of each solution. The focus of this brainstorming session is on the perceived improvements that the solution is likely to bring about as they would be perceived by the organization's customers.

- Identify organizational consequences of each solution.

  Conduct a team brainstorming session to identify qualitative solutions' impacts as they pertain to the organization internally. These impacts would affect "quality-of-life" considerations (e.g., rewards and compensation, work environment, etc.), morale, etc.
- Identify process-related consequences of each solution.

  Conduct a team brainstorming session to identify quantitative impacts of each solution. These impacts include effects on cycle time, costs, etc. Costs should be expressed in the organization's preferred format for cost/benefit analyses (e.g., internal rate of return, net present value, etc.).
- Evaluate the solutions.

  Evaluate each solution given all the qualitative and quantitative impacts of its implementation. Identify solutions for further consideration.

Example

The organization's staff was having problems with a pressure valve they were producing and selling. They were producing 10,000 per month and believed that there will be a demand at that level for the next 30 months if the problem is solved. The cost to replace a new valve that had been delivered to the customer is $35. Projected field failures are 300,000 ÷ 230 = 1,304. Cost for field failures 1304 × $35 = $45,640. The team defined three potential solutions to the problem they were addressing. They included:

1. Change the file test program
2. Rework the holding fixture
3. Change the engineering design

The following is their impact solution evaluation.

- Solution 1
  - Customer impact: Reduce the failure rate down from 1 in 230 to 1 in 1,500.
  - Process impact: Increase processing time by seven minutes and increase the defects.

- Solution 2
  - Customer impact: Reduce a failure rate down from 1 in 230 to 1 in 1,800.
  - Process impact: Increased manufacturing costs by $0.15 per unit.
- Solution 3
  - Customer impact: Eliminated the impact on the customer.
  - Process impact: Increased manufacturing costs by $ 0.10 per unit, reduced processing time by 20 minutes.

The following is the resource solution evaluation:

- Solution 1
  - Implementation costs: $1,800
  - Cycle time to design and implement: 14 days
  - Risks related to the solution: low
- Solution 2
  - Implementation costs: $400
  - Cycle time to design and implement: 14 days
  - Risks related to the solution: medium
- Solution 3
  - Implementation costs: $3,500
  - Cycle time to design and implement: 28 days
  - Risk related to the solution: low

A quick analysis indicated to the team that Solution 2 or Solution 3 were both better than Solution 1. In reviewing the cost benefits analysis for Solution 2 and Solution 3, the team selected Solution 3, even though it cost more to implement and took additional time, because the overall savings were greater and the solution gave a higher level of customer satisfaction.

40

Storyboarding

DEFINITION

Storyboarding is physically structuring the output into a logical arrangement. The ideas, observations, or solutions may be grouped visually according to shared characteristics, dependencies upon one another, or similar means. These groupings show relationships between ideas and provide a starting point for action plans and implementation sequences.

Its Uses

Storyboarding may be used to give structure to data, ideas, information, observations, etc.

General

Storyboarding coordinates "ideas" or "solutions," thereby demonstrating its application in a problem-solving context. However, it should be noted that storyboarding is an effective technique in many situations and you should consider its use in other contexts.

Storyboarding helps put items being discussed in focus and provides for building an understanding of the particular problem or issue. Storyboarding helps break down an activity or problem into more of a visual scene, which, in turn, provides a clearer picture of what is really going on. Storyboarding also can be used to help define a situation or a solution.

The output of a storyboarding session is a sequence of events/activities and/or concepts that flow in a logical order and build upon a problem, issue, or idea. A number of different types of resource material may be used in this technique, including:

- Flip-chart
- Post-it® notes
- 3" × 5" cards
- 5" × 7" cards
- Whiteboard
- Plain paper

The desired outcome is to show each step and its relationship to the next in order to form a clear picture of the "problem" or situation (Figure 40.1).

A storyboard is normally made out of a 4 ft. × 8 ft. sheet of corkboard, whiteboard, or a large piece of butcher-type paper. You would normally pin or tape index cards and other things to the board to tell the story about the problem or issue on which you are working. There are four types of storyboards most often used:

- Planning
- Idea
- Communication
- Organization

Preparation

- Record solutions on index cards or another type of note paper:
    - If solutions have already been developed, transfer each one onto a self-adhesive note or an index card. Otherwise, instruct each

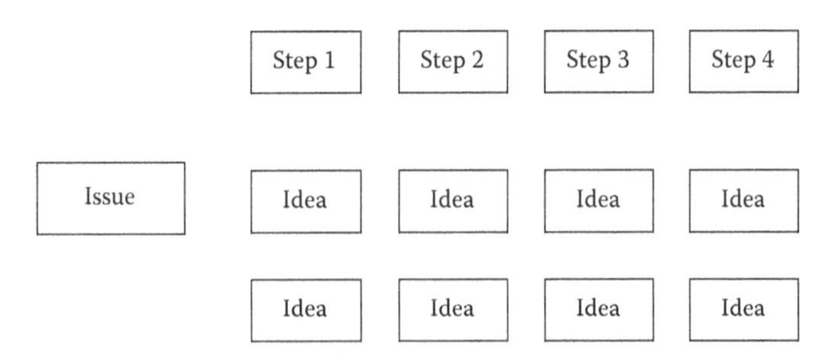

FIGURE 40.1
Storyboard layout.

member of the group to write his or her ideas onto individual notes or cards. Write only one idea on each self-adhesive note or card.
    - If the group dynamic provides a constructive work environment, you may wish to modify this step by incorporating elements of brainstorming. This could be accomplished by having each participant share ideas with the group as they occur, thereby creating opportunities to build on ideas.
- Display the ideas:
    - Post all ideas on a board in random order. The board should be fairly large. Discuss ways to group the notes.
    - Arrange notes in predefined groups. The group may wish to employ one of them, combine elements of them, or create a scheme of your own. Your decision will be based on the nature of the issue and whether it is best presented in a flow-like format, according to shared characteristics, etc.
- Create the storyboards:
    - Create the storyboard using either the Network, Cluster, Matrix, or Tree Structure format. See the following:
        - Network: Review each idea for similarities or relationships to other ideas. Draw a line between sets of connected or related ideas.
        - Cluster: Develop a set of category headings (the number of these varies according to the number of ideas that are being considered; two to six headings is a typical range). The category headings are based on major themes that emerge in the course of developing ideas. Attach the headings across the upper edge of the storyboard. Rearrange the notes or cards by posting them below the appropriate headings. This process will likely generate discussion around ideas that could fall under two or more categories. For resolution of these issues, consider placing the card between two headings, if possible, or apply consensus building or voting rules as appropriate. If a sufficiently large number of ideas fall under multiple categories, it may be more appropriate to use the matrix structure described below.
        - Matrix: Develop a list of categories. List all category headings vertically and horizontally. Post ideas according to their fit with the vertically and horizontally listed headings.

- Tree Structure: Review ideas and determine which ones must occur before others can be implemented. Arrange cards or notes according to the sequence in which they should occur. The board should flow from left to right. Therefore, begin posting those ideas that must be implemented ahead of others on the left side of the board.

Example

See Figure 40.2 and Figure 40.3.

Let's pretend we own an ice cream parlor and want to know what our improvement opportunities are from the customer's point of view. We would start by deciding what type of storyboard to use. Because our view of what the customer sees as our improvement opportunities is conceptual, we would use an *Idea* storyboard to give us a picture of the customer's view.

Step 1. We would start by writing out the task, "Improvement Opportunities from the Customers' Point of View." This would be posted to the left side of our board.

Step 2. Next we would determine what "categories" the customer might be interested in. They could include:
- How easy are we to find?
- How convenient is our parking?
- Is our building accessible?

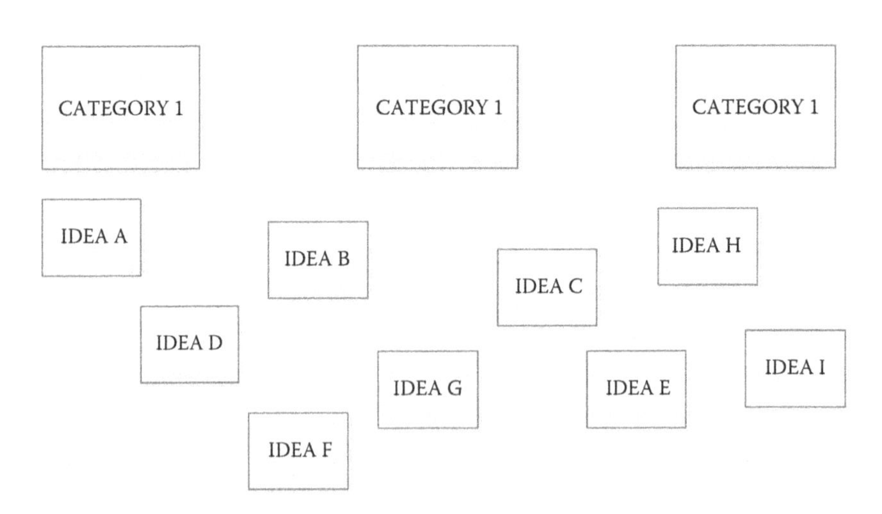

FIGURE 40.2
Starting layout for the storyboard.

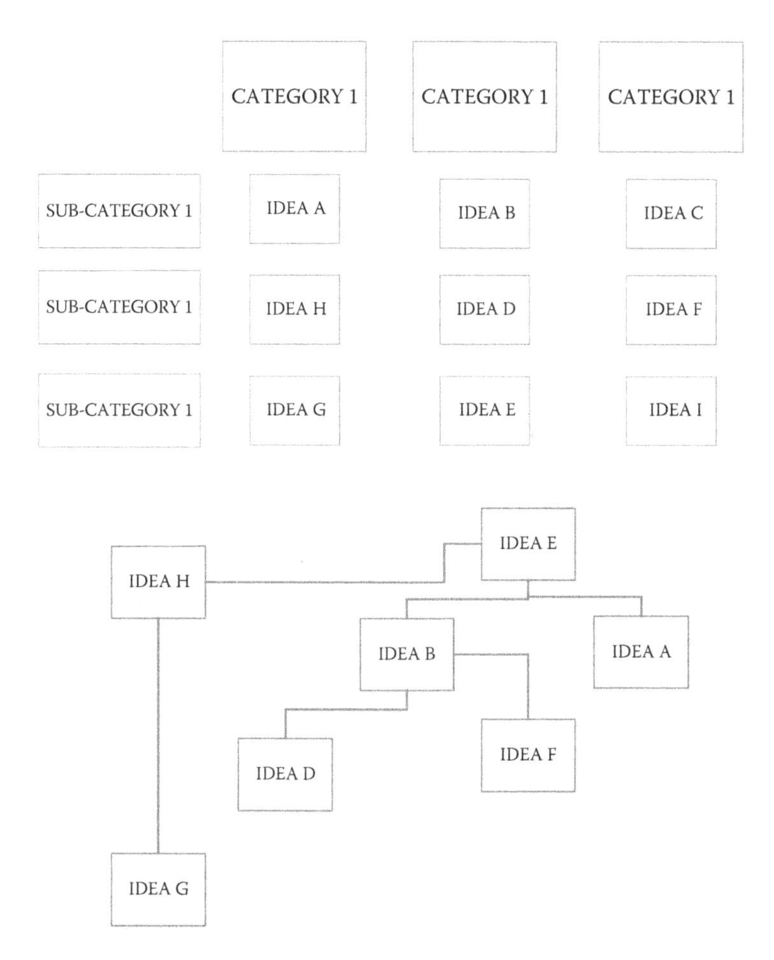

FIGURE 40.3
Simple storyboard.

- Is our signage easy to read?
- How about our workers?
- How good is our seating?
- How is the product/service?

Step 3. Each of the categories would be posted along the top of our board.

Step 4. For our first category (How easy are we to find?), we could go outside and take a picture of how we are viewed from the road. Under the category, we could place "issues," such as:

- How well can customers view our building?
- How well can customers view our signage?

For each additional category we would follow the same procedure, looking for issues the customer might view as potential improvement opportunities.

After our storyboard is complete, we would review our data and test the validity of each potential improvement idea. Those that are valid we would correct.

Software

Some commercial software available includes, but is not limited to:

- StoryBoard Quick 6™
- SmartDraw®
- StoryBoard Quick™
- StoryBoard Creator™

ADDITIONAL READING

Cartin, T. J. 1993. *Principles and practices of TQM.* New York: McGraw-Hill.

Ernst & Young Quality Improvement Group. 1992. *Tools and techniques resource guide.* San Jose, CA: Ernst & Young LLP.

Forsha, H. I. 1995. *Show me: The complete guide to storyboarding and problem solving.* Milwaukee, WI: ASQ Quality Press.

Tague, N. R. 1995. *The quality toolbox.* Milwaukee, WI: ASQ Quality Press.

41

Value Analysis

DEFINITION

Value analysis (VA) is the act of identifying the required functions for a product, establishing values for the required functions, and suggesting an approach to provide the required functions at the lowest overall cost without performance loss to optimize cost performance.

Its Uses

The objective of value analysis is to optimize real value-added activities and minimize or eliminate no value-added activities/tasks.

General

Value is defined from the point of view of the external customer. There are three classifications of the value activities. They include:

1. Real value-added (RVA) activities. These are activities that, when viewed by the external customer, are required to provide the output that the customer is expecting.
2. Business value-added (BVA) activities. These are activities that need to be performed in order to run the organization, but add no value from the external customer's standpoint (e.g., updating operating procedures, filling out timecards, employee performance reviews, etc.).
3. No value-added (NVA) activities. These are activities that do not contribute to meeting external customer requirements and could be eliminated without degrading the product, service, or the business. Typical examples would be checks and balances in the organization rework, inspecting parts, rewriting reports, or engineering changes.

In most organizations there is a huge hidden office made up of BVA and NVA activities that often exceed the cost of the RVA activities. The challenge that faces most organizations is to minimize the BVA activities and to eliminate the NVA activities.

Preparation

Based on our experience, over 90 percent of all the processes used within an organization contain some measure of NVA and/or BVA content. As a result, almost every process and system within your organization is a good candidate for value analysis. We recommend starting with processes that consume a great deal of resources or are causing problems and delays within the organization. Another good approach that is frequently used by organizations is focusing first on the processes that interface directly with the external customer.

Often this tool is used in conjunction with a major methodology like Business Process Improvement, Total Quality Management, Six Sigma, Activity-Based Costing, etc. In these cases, the process that the value analysis tool will be applied to has already been selected as part of the larger initiative.

Once the process that the value analysis tool will be applied to has been selected, the following activities will be taken in order to effectively complete the analysis.

- A team is formed that is made up of at least one individual who is familiar with the value analysis tool, and individuals who are familiar with the process that the team is assigned to evaluate. This team can be as small as one and should not exceed 10 members.
- The team members are trained in using the value analysis tool.
- The team should determine the impact that eliminating the process would have upon the organization's performance. Is the process really needed? If not, it should just be eliminated. With the elimination of the process, the project is completed.
- A flowchart of the process is prepared for each activity and classified as being RVA, BVA, or NVA. It is important to note that, in some cases, an individual activity can contain parts of it that are made up of all three categories. Some individuals prefer preparing a value stream map in place of a flowchart. Both work equally well.

- The team should collect cycle time, processing time, wait time costs, and quality data by doing a process walk-through for each activity. This information is typically added to the flowchart. With the addition of this information, the present state map of the process is complete.
- The team then develops action plans to eliminate the NVA activities. (Particular emphasis is placed on eliminating bureaucracy, movement of parts and paper, wait times, inspection, scrap, and rework.)
- The team then will focus on reducing, simplifying, or eliminating BVA activities. (Can it be simplified? Can it be done automatically? Is it really necessary? Is it a duplication of previous effort? Is all the information that was collected going to be used, etc.)
- The team then creates a future state map of the process that is used to document the proposed changes to the process and to estimate the projected performance improvements.
- The proposed future state process then is piloted along with the control sample to validate the projected performance improvement and to ensure that the future state process did not create unexpected performance problems. In some cases, when the changes to the process are minimal and very obvious, it may not be necessary to run a pilot analysis.
- The future state process should then be documented to the level required based on the skills of the employees that will be using the process.
- The employees are introduced and trained on how to use the future state process and the process is implemented.
- Measurement systems are put in place to measure the performance improvement of the future state process, which is now the current state process, to validate the projected performance improvement and to ensure additional problems were not created.

Example

See Table 41.1.

By reviewing Table 41.1, you can see that the only part of the activity that is RVA is the part of the conversation when you provided the customer with a date of delivery. This was about five seconds of the conversation when you said, "Your order 9072 for 8,000 brass rivets will be shipped on January 14 by Federal Express Mail." The rest of the process was BVA or NVA. You will

TABLE 41.1

Tasks to Contact Customer for Delivery Status

Turn on your computer	No value-added
Search through the inputs until you find the customer order status	No value-added
Read the status to be sure it covers all that the customer ordered	Business value-added
Find your scheduling notebook	No value-added
Record the data in your notebook	Business value-added
Look up the customer's phone number	No value-added
Place the call to the customer	No value-added
Customer calls back	No value-added
Find your scheduling notebook and find the status of the order	No value-added
Provide the customer with the date the order will be delivered	Real value-added
Record in your scheduling notebook that you have completed the task	Business value-added
Put your scheduling notebook away	No value-added

see from this example that even the processes that are classified as RVA activities have significant opportunity for improvement. Most of the activities classified as RVA activities have less than 20 percent of the costs devoted to RVA activities.

In this case, the real future state solution should be to establish a portal so the customer can inquire directly related to his/her order status and projected delivery dates.

Software

Some commercial software available includes, but is not limited to:

- Visio 14.0.6
- SmartDraw® VP 19.1.3.2
- Flowcharter® 14.1.2
- Edraw Flowchart 6
- EDGE Diagrammer 6.24
- WizFlow Flowcharter 6.24
- RFFlow 5.06

42

Value Engineering (VE)

DEFINITION

Value engineering (VE) is a methodology that seeks to improve the "value" of goods or products and services by evaluating ways that costs or function can be improved without having a negative impact upon the other parameter. Value can be increased by either improving the function of the item or reducing the cost to produce the item. Value, as defined, is the (function)/(cost). Therefore, improving the function or reducing the cost will increase the value. Basic functions must be preserved and not reduced as a consequence of pursuing value improvements. In the United States, value engineering is specifically spelled out in Public Law 104-106, which states, "Each executive agency shall establish and maintain cost-effective value engineering procedures and processes."

Its Uses

It is a methodology used to decrease the cost of a product or process without having a negative impact upon the item's performance as viewed by the intended consumer or to increase the performance as viewed by the consumer without increasing costs of the item.

General

Costs can be reduced by simplifying processes in manufacturing or service or by using fewer parts or less expensive parts and materials. Great care should be used when implementing this methodology to be sure that the performance of the item is not impacted negatively. Every process step and material used should add value and yield a product that gives the best performance.

Preparation

Depending on the application, there may be four, five, six, or more stages. One modern version has the following seven steps:

1. Preparation: Identify the problem or situation
2. Information: Identify current and improvement requirements
3. Analysis: Evaluate effectiveness of current processes
4. Creation: Create an alternative process or materials
5. Development: Develop an implementation plan
6. Evaluation: Evaluate effectiveness of proposed changes
7. Follow-up: Control the improvements

Example

- If you have ever stayed at a Travelodge hotel/motel, you might have noticed there isn't shampoo in the bathroom.

  VE identified its customers' needs.

  "Pay for things you don't want? That's crazy! Our research shows that most people staying in a hotel simply want a clean, comfortable place to get a good night's sleep, and are happy to forgo the unnecessary "frills" offered in other stuffy, over-priced establishments. So, we make sure we provide good quality essentials, such as a comfortable bed and a decent quality shower, but get rid of unnecessary extras." (See www.bbc.co.uk/blogs/legacy/thereporters/evandavis/2007/05/value_engineering.html)
- From Cadcam-e.com (CCE):

  One of CCE's customers is a furniture manufacturer. They have a round base that is used as base support in chairs and bar stools. The customer wanted CCE to investigate if there were opportunities to reduce materials used in manufacturing the round base without affecting design intent or performance. They provided CCE with 3D CAD models and 2D drawings of the part.

  CCE engineers studied the functional application of the round base and how it was being used at the base of chairs and stools. CCE engineers redesigned the part by removing excess material from the round base that was not adding any structural value. Special care was taken to avoid any impact on the aesthetics of the part. CCE's new design had a weight of 15.5 kgs when compared to the original weight of 20.5 kgs.

Key highlights of this project include:

- Component weight reduction: 25 percent.
- Reduction in material: Component weight reduced 25 percent from 20.5 kgs to 15.5 kgs.
- Safer design: Displacement max values were higher in the new design as compared to the original.
- Better balance: Center of gravity shift is minimum, and that, too, is only in the Y-axis, which is adjustable.

CCE's design was aesthetically superior to existing design based on customer feedback.

Proposal very well received by the customer; design has been prototyped for manufacturing. (See www.cadcam-e.com/pdf/CCE Case Study-Value Engineering (Material Reduction) of Round Base.pdf) Software

Some commercial software available includes, but is not limited to:

- Edraw max
- SmartDraw®
- Affinity Diagram 2.1
- QI Macros™

ADDITIONAL READING

Cooper, R., and R. Slagmulder. 1997. *Target costing and value engineering*. Portland, OR: Productivity Press.

Mukhopadhyaya, A. K. 2009. *Value engineering: Concepts, techniques and applications*. Thousand Oaks, CA: SAGE Publications.

Mukhopadhyaya, A. K. 2009. *Value engineering mastermind: From concept to value engineering certification*. Thousand Oaks, CA: SAGE Publications.

Stewart, R. B. 2010. *Value optimization for project and performance management, CVS-Life, FSAVE, PMP*, 1st ed. Hoboken, NJ: John Wiley & Sons.

43

Value Proposition

DEFINITION

A value proposition is a document that defines the benefits that will result from the implementation of a change or the use of an output as viewed by one or more of the organization's stakeholders. A value proposition can apply to an entire organization, parts thereof, or customers, or products, or services, or internal processes.

A business case captures the reasoning for initiating a project or task. It most often is presented in a well-structured, written document, but, in some cases, it may come in the form of a short, verbal agreement or presentation. A business case is preferably prepared by an independent group after the project concept has been approved by executive management (Harrington and Trusko, 2014).

Its Uses

This tool can be used by just one individual, but its best use is with a group of four to eight people. Usually, the best results from preparing a value proposition are realized when the group is made up of members from cross-functional units. The value proposition is extensively used in the phases that are followed with an X. It is not used extensively in the other listed phases.

- Creation phase ___
- Value proposition phase _X_
- Financing phase _X_
- Documentation phase __
- Production phase __
- Sales/delivery phase __
- Performance analysis phase __

General

Value propositions are prepared to determine if the proposed project will have a positive or negative value-added content to the major stakeholders. It takes into consideration all of the costs related to managing and implementing the project, the cost of operating the affected systems/processes compared to the value added in continuing to operate without making any changes. In most cases, these are only estimates. As a result, the organization needs to seriously consider how accurate it is in making a decision to approve or disapprove the proposed project.

A costs benefits analysis (CBA) is a critical part of a value proposition, and is defined as anything used to compare the benefit of a proposed process change with the cost associated with implementing that change. The higher the ratio of the measurable benefits to the quantifiable costs, the easier it is to spend resources on a proposed process implementation. In the simplest form of CBA, there are two primary measures:

1. Positive financial benefit = Financial impact – The costs of implementation
2. Payback = Implementation costs/annual benefits

When doing a CBA, there are some key principles that need to be considered:

- Use the same units of measurement for both costs and benefits.
- Compare the situation with and without the process change to assess its real impact.
- Consider the specific location of the study (and its unique characteristics) for transferring the results to the other area.
- Avoid double counting either benefit for costs.

Preparation

Steps in preparing a value proposition:

1. Define the costs related to the value proposition group's activities.
2. Estimate the cost related to implementing the proposed project.
3. Estimate the time required to implement the proposed project.

4. Define the risks related to implementing the proposed project. Define both the positive and negative impacts the project will have on major stakeholders (both tangible and intangible).
5. Define assumptions upon which the estimates are based.
6. Prepare the value proposition report.
7. Present the value proposition report to executive management.

Example

The following is a typical example of the Table of Contents for a typical value proposition report.

- Executive Overview
- Description of Current State
- Added Value That the Proposed Change Would Produce
- Description of the Proposed Change
- Backup Data
- Costs and Required Timeframe
- Other Solutions Considered
- Risk and Obstacles
- Recommendations
- Key Individuals
- Financial Calculations
- Other Value-Added Results: Both Tangible and Intangible
- Lists of Assumptions
- Implementation Plan

Note: Be sure that you make your point in the executive overview. The rest of the report only serves as a backup. Some executives will not have the time to read the full report and will make their decision based on the executive overview.

Software

Some commercial software available includes, but is not limited to:

- WinSite®
- Logix Guru

REFERENCE

Harrington, H. J., and B. Trusko. 2014. *Maximizing your value proposition.* Boca Raton, FL: Taylor & Francis Group.

ADDITIONAL READING

Gyorffy, L., and L. Friedman. 2012. *Creating value with CO-STAR: An innovation tool for perfecting and pitching your brilliant idea.* Palo Alto, CA: Enterprise Development Group.

Hardy, J. G. 2005. *The core value proposition: Capture the power of your business building ideas.* Cheshire, U.K.: Trafford Publishing.

Palomino, J. 2008. *Value prop: Create powerful 13 value propositions to enter and win new markets.* Philadelphia: Cody Rock Press.

44

Voting

DEFINITION

Voting is a method of selecting an item from a list of alternatives. It is a method that allows an individual to identify his or her preference from a group of alternatives.

Its Uses

Voting is most effective when consensus is difficult to reach.

General

The technique can be performed by secret ballot or by a show of hands. The advantage of voting is its speed, making it most effective when consensus is difficult to reach. The disadvantage is that it may create an "us and them" atmosphere. For this reason, the technique is most effective when used to narrow a list of alternatives, rather than come up with a final decision.

Preparation

Ballot Voting

- Determine the number of items from the list of alternatives that each participant is to identify.
- Agree upon the criteria on which each alternative is to be evaluated.
- Individually rank the agreed-upon number of items. The item with the highest priority should have the highest valued numeric ranking. Therefore, the lowest priority item will have a rank of 1.
- Record all individual rankings on a master list.

- Tally the numeric values given to each alternative. Identify the alternatives receiving:
    - Highest total value
    - Highest number of mentions
    - Top priority by any individual
- Further discuss the list of alternatives. Further voting may occur if necessary.

Show-of-Hands Voting

The procedure is similar, but this method is typically used when the group wants to narrow down to one alternative. Individuals vote for only one alternative instead of rank ordering several alternatives.

What Is a Voting Output?

A voting output is an alternative or narrowed list of alternatives on which further decisions can be made.

Voting

Prepare the Voting

- Determine the number of items from the list of alternatives that each participant has identified.
- Agree upon the criteria on which each alternative is to be evaluated. For example, personal preference, cost, resistance, time, resources, or risk.

Conduct the Voting

- Individually rank the agreed-upon number of items. The item with the highest priority should have the highest valued numeric ranking. Therefore, the lowest priority item will have a rank of 1.
- Record all individual rankings on a master list.
- Total the numeric values given to each alternative.
- Identify the alternatives receiving:
    - Highest total value
    - Highest number of mentions
    - Top priority by any individual
- Discuss the list of alternatives. Vote again if necessary.

Example

A team of 10 people were assigned to improve the accounts payable process. After discussing how to go about improving the process, part of the group was split between two methodologies. Part of the group wanted to use a process reengineering approach and part of the group wanted to use a process redesign approach. After both sides stated why their particular approach should be used, the team leader called for a vote. Three people voted for using a process redesign approach, five people voted for using a process reengineering approach and two people did not vote for either approach. As a result, the team developed their project plan based on the process redesign methodology. Why did they decide to use the process redesign approach? The team had additional discussions where they identified the risk, the cost, and cycle time related to the two approaches. After qualifying the risk, cycle time, and the costs, the team decided that the risk, costs, and time to reengineer was not justifiable when compared to the same elements in the process redesign methodology. You will note that everyone did not vote for the same methodology. It is essential that everyone in the team agrees to adopt the methodology that the majority of the team believes is correct.

Software

Some commercial software available includes, but is not limited to:

- SourceForge
- Capterra®

45

Workflow Diagram

DEFINITION

A workflow diagram visually represents the movement and transfer of resources, documents, data, and tasks through the entire work process for a given product or service. The diagram is based on the layout of the organization or the department with which the process comes in contact.

Its Uses

Workflow diagrams are used to minimize the time, distance, and cost related to processing an item through one of the organization's processes.

General

It makes use of a drawing of the physical layout of a department and/or work area that an item comes in contact with. Arrows are used to show how the item moves from one place to another.

This layout is studied and restructured to reduce the different since traveled, cycle time, and the cost related to processing.

Preparation

Following are typical activities that the Process Improvement Team (PIT) will go through in order to create a workflow diagram and use it to improve the related process.

- Management selects the process that has an opportunity to improve its overall performance and it assigns a team to undertake the study.

- The PIT defines the starting and ending boundary and prepares a block diagram of its understanding on how the process produces an output.
- The PIT will then conduct a process walk-through where the members physically observe how the process functions. During this walk-through, part of the information they collect will be used to construct the workflow diagram. The type of information that will be needed is the physical location in the organization where the activity is being conducted, the distance that the work product moves between activities, the cycle time expended between the time that the last activity was physically being performed and the time that the next activity physically starts, the way that the work product moves from one activity to the next, and the cost of moving the work product from activity to activity. This information is then laid out on an Excel® spreadsheet.
- Now the PIT will construct a block diagram locating the process activities in a relative position to each other as physically laid out within the organization. The movement of output between activities is then indicated by arrows that run from activity to activity. This is commonly referred to as a workflow diagram or a workflow map.
- The PIT then brainstorms to define ways to minimize the cycle time, movement of output, and cost related to the process. Particular focus is on reassigning and combining activities, changing the location that activities are performed at and changing the order that activities are performed.
- Based on an analysis of the brainstorming session, a discussion of options for a future state layout is developed.
- The PIT then prepares a workflow diagram and an Excel spreadsheet of the proposed future state layout.
- The PIT presents a comparison of the present state process to the proposed future state process to the appropriate management team to get their approval or disapproval for implementation.
- Based upon the estimated performance improvements, a decision is made by management related to implementing the proposed future state layout.

Example

Figure 45.1 is a workflow diagram of the first day process a new employee follows when he/she reports to work at one of our client's locations.

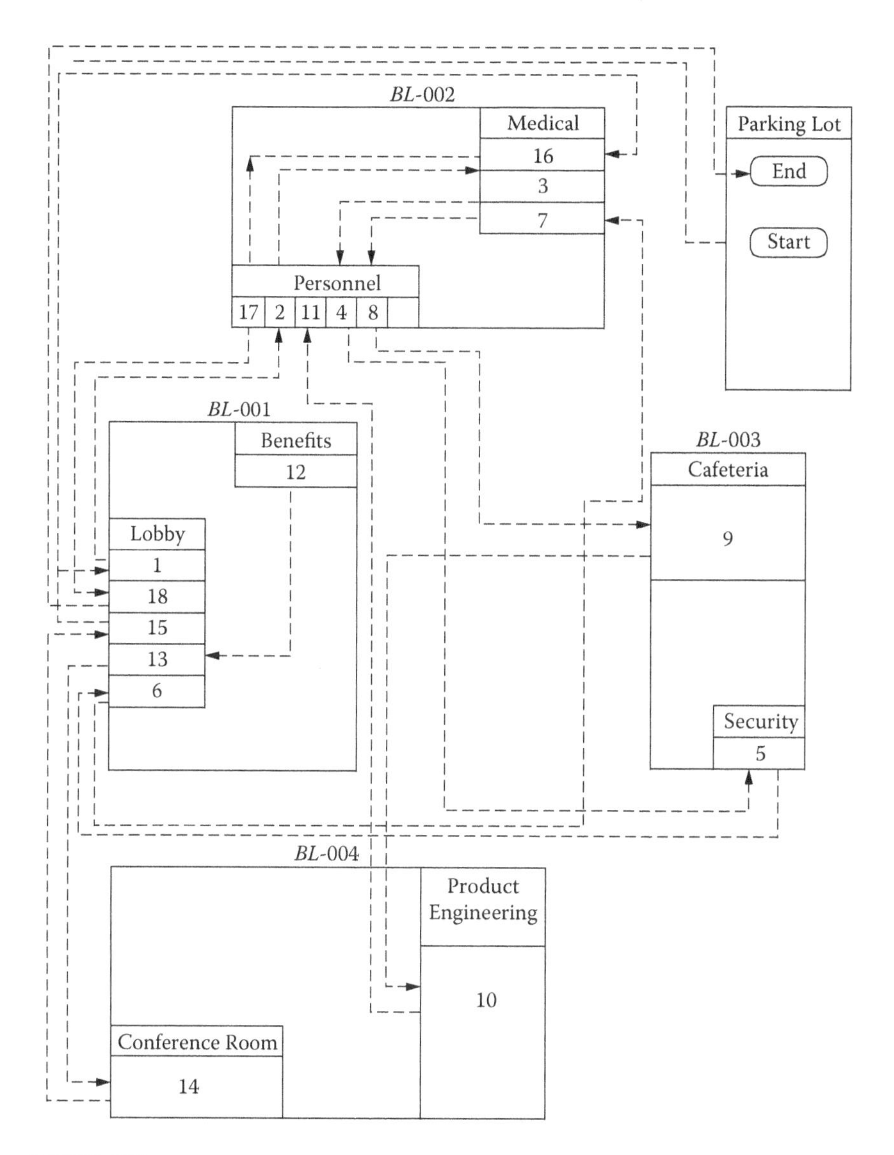

FIGURE 45.1
Workflow diagram for an individual's first day with a new organization.

The following is a list of the activities the new employee is involved with in order to start work on his/her first day with a new organization:

- New employee signs in at the lobby and asks the receptionist to call Personnel.
- Personnel receptionist greets new employee and takes him/her to the Personnel Department to review pertinent procedures.

- Placement representative takes new employee to medical department to fill out medical forms and makes an appointment with the nurse for required tests.
- New employee returns to Personnel Department to fill out payroll forms.
- New employee and placement representative go to security to have his/her picture taken and to obtain a temporary identification badge.
- New employee returns to lobby to wait for appointment with nurse. He/she can go unescorted now that he/she has a temporary badge.
- New employee goes to medical department for blood tests and makes an appointment for a physical exam with the doctor.
- New employee returns to Personnel Department for instructions.
- Placement representative takes new employee to lunch.
- Placement representative takes new employee to meet his/her new manager and tours the department.
- New employee goes to Personnel Department so that personnel can take him/her to Benefits Department.
- New employee reviews benefit package and selects a benefit plan.
- New employee goes to lobby to wait for new employee orientation meeting.
- New employee attends new employee orientation meeting.
- New employee returns to lobby to wait for appointment with doctor.
- New employee goes to medical department for appointment with doctor and returns to Personnel Department.
- Personnel reviews new employee checklist and calls medical to find out whether exam results were favorable.
- New employee returns to lobby, turns in temporary badge, and signs out.

Figure 45.2 is a future state workflow diagram of the first day process a new employee follows when he/she reports to work after the PIT has streamlined the process.

Figure 45.3 is a workflow diagram of the process to investigate a customer claim.

Figure 45.4 is a future state workflow diagram of the process to investigate a customer claim after the PIT has streamlined the process.

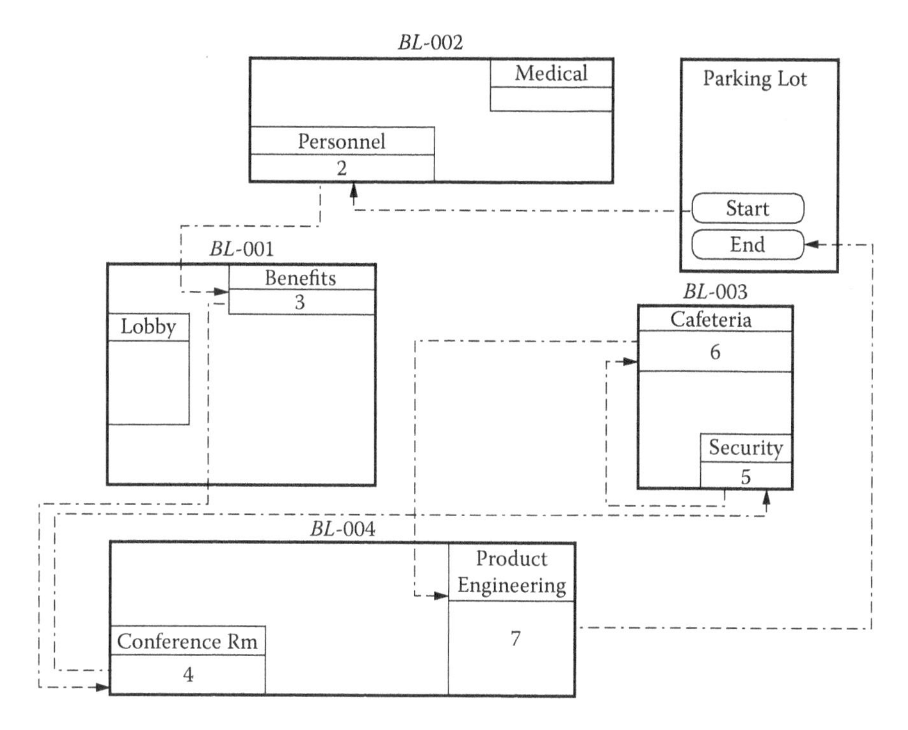

FIGURE 45.2
Future state workflow diagram of the first day process a new employee follows after the workflow had been streamlined.

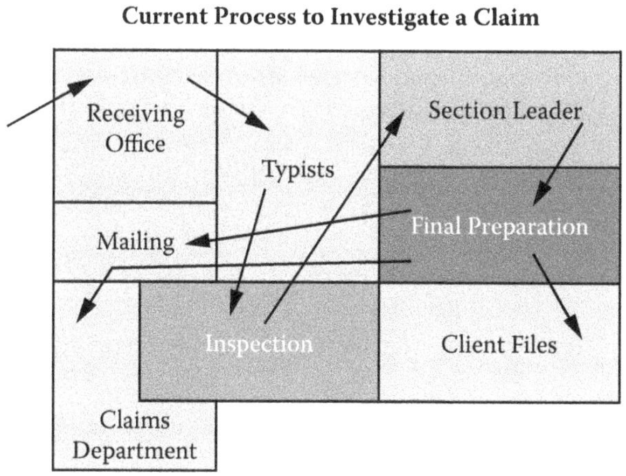

Current Process to Investigate a Claim

Receiving Office

Typists

Section Leader

Mailing

Final Preparation

Inspection

Client Files

Claims Department

FIGURE 45.3
Present state workflow diagram of a process to investigate a claim.

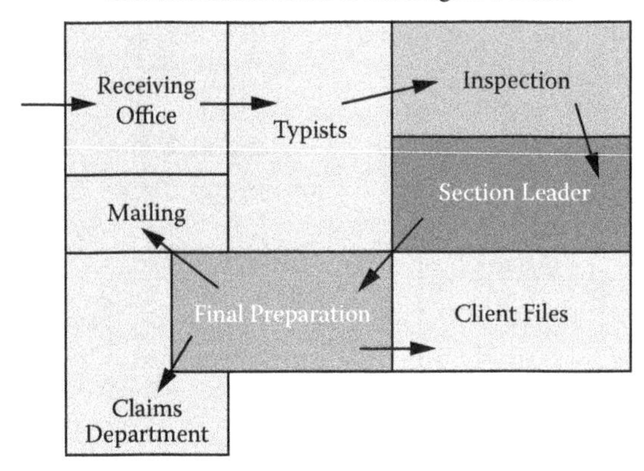

FIGURE 45.4
Future state of a process to investigate a claim after it had been streamlined.

Software

Some commercial software available includes, but is not limited to:

- Intuit Quick Base®
- Edraw Flowchart
- Grapholite

ADDITIONAL READING

Harrington, H. J. 1991. *Business process improvement.* New York: McGraw-Hill.
Harrington, H. J. 2012. *Streamlined process improvement.* New York: McGraw-Hill.

Glossary

5Ws and 2Hs: A rigid, structured approach that probes into and defines a problem by asking a specific set of questions related to a previously defined opportunity or problem statement. The 5Ws and 2Hs stands for

- W1: What?
- W2: Why?
- W3: Where?
- W4: Who?
- W5: When?
- H1: How did it happen?
- H2: How much did it cost?

Activity-Based Costing (ABC): A technique for accumulating product cost by determining all costs associated with the activities required to produce the output.

Activity Plan: A simple chart that shows a list of implementation activities listed in sequence. It identifies the individual responsible for a particular activity and the projected timing of that activity.

Adaptability: The flexibility of a process to handle future, changing customer expectations and today's individual, special customer requirements. It is managing the process to meet today's special needs and future requirements. Adaptability is an area largely ignored, but is critical for gaining a competitive advantage in the marketplace. Customers always remember how you handle or don't handle their special needs.

Advocate: An individual/group who wants to achieve change, but does not have sufficient sponsorship.

Affinity Diagrams: A technique for organizing a variety of subjective data (such as options) into categories based on the intuitive relationships among individual pieces of information. Often used to find commonalities among concerns and ideas.

Appraisal Costs: These are the costs that result from evaluating already completed output and auditing the process to measure compliance to established criteria and procedures. To say it another way, appraisal costs are all the costs expended to determine if an activity was done right every time.

Area Activity Analysis (AAA): A proven approach used by each natural work team (area) to establish efficiency and effectiveness measurement systems, performance standards, improvement goals, and feedback systems that are aligned with the organization's objectives and understood by the employees involved.

Arrow Diagrams: A way to define the most effective sequence of events and control the activity in order to meet a specific objective in a minimum amount of time. It is an adaptation of PERT (Program Evaluation and Review Technique) or the CPM (Critical Path Method).

Ask "Why" 5 Times: A systematic technique used to search for, and identify the root cause of a problem.

Assumption Evaluation: Provides a way of redefining problem statements, analyzing solutions, and for generating new ideas.

Attribute Control Chart: A plot of attributes data of some parameter of a process's performance, usually determined by regular sampling of the product, service, or process as a function (usually) of time or unit number or other chronological variables. This is a frequency distribution plotted continuously over time, which gives immediate feedback about the behavior of a process. A control chart will have the following elements:

- Center line (CL)
- Upper control limit (UCL)
- Lower control limit (LCL)

Attributes Data: This is counted data that can be classified as either yes/no, accept/reject, black/white, or go/no-go. These data are usually easy to collect because they require only counting and are not measuring the process, but they often require large samples.

Bar Graph: Have bands positioned horizontally (bars) or vertically (columns) that, by their height or length, show variations in the magnitude of several measurements. The bars and columns may be multiple to show two or more related measurements in several situations.

Bell-Shaped Curve: The shape of a normal distribution curve.

Benchmark: A reference point where other items can be compared. It can be a location, a process, a measurement, or a result.

Benchmarking (BMKG): A systematic way to identify, understand, and creatively evolve superior products, services, design, equipment, processes, and practices to improve your organization's real performance.

Best Practice: A process or a method that is superior to all other known methods.

Best-Value Future-State Solution: A solution that results in the most beneficial new item as viewed by the item's stakeholders. It is the best combination of implementation cost, implementation cycle time, risk, and performance results (examples: return-on-investment, customer satisfaction, market share, risk, value-added per employee, time to implement, cost to implement, etc.)

Black Belts: They are highly trained team leaders responsible for implementing process improvement projects within an organization. They are normally full-time assignments, for each Black Belt is expected to save the organization a minimum of $1 million per year. They have been trained to manage the projects by fact, not by gut feel. Black Belts coach Green Belts and receive coaching support from Master Black Belts.

Block Diagrams: A pictorial method of showing activity flow through a process, using rectangles connected by a line with an arrow at the end of the line indicating direction of flow. A short phrase describing the activity is recorded in each rectangle.

Brainstorming (Creative Brainstorming): A technique used by a group to quickly generate large lists of ideas, problems, or issues. The emphasis is on quantity of ideas, not quality.

Bureaucracy Elimination Method: An approach to identify and eliminate checks and balances activities that are not cost justified.

Business Case Development: An evaluation of the potential impact a problem has on the organization to determine if it is worthwhile investing the resources to correct the problem or take advantage of the opportunity.

Business Objective: Defines what the organization wishes to accomplish over the next 5 to 10 years.

Business Process Improvement (BPI): This is a breakthrough methodology that includes process redesign, process reengineering, process benchmarking, and fast action solution teams.

Cause-and-Effect Diagrams: A visual presentation of possible causes of a specific problem or condition. The effect is listed on the right-hand side and the causes take the shape of fish bones. This is the reason it is sometimes called a "Fishbone Diagram." It is also called "Ishikawa Diagram."

Central Tendency: A measure of the center of the distribution.

Change: Individual/group who must actually change. A changee is also called a change target.

Change Agent: Individual/group who is responsible for implementing the change.

Check Sheet: A simple form on which data are recorded in a uniform manner. The forms are used to minimize the risk of errors and to facilitate the organized collection and analysis of data.

Commitment Building: Commitment is a promise to give or do something, to be loyal to someone or something. It is the act of pledging or engaging oneself in an obligation or a promise to be engaged, or becoming involved in a given activity to achieve a given result.

Common Cause: A source of errors that is always present because it is part of the random variation in the process itself. These types of failures are normally traced back to the process that only management can correct.

Comparative Analysis: A systematic way of comparing an item to another item to identify improvement opportunities and/or gaps. (It is the first three phases in the benchmarking process.)

Competitive Benchmarking: A form of external benchmarking that requires investigating a competitor's products, services, and processes. The most common way to do this is to purchase competitive products and services and analyze them to identify competitive advantages.

Confidence Limits: This is a calculated measure of the accuracy of the results obtained from pulling a sample of a complete population. For example, your confidence level may be 90 percent that the cycle time is three hours plus or minus 2 percent.

Consensus: An interactive process, involving all group members, where ideas are openly exchanged and discussed until all group members accept and support a decision, even though some of the groups' members may not completely agree with it. To reach a consensus after is time consuming and often involves individual compromising.

Consensus Building: A technique to obtain the commitment of all team members to move in a particular direction.

Consequence Management: A formal process of understanding the institutional structures that reflect deeply held values and beliefs in the organization and then utilizing those structures (e.g., compensation or training opportunities) to influence desired behavior.

Control Charts: A graphic representation that monitors changes that occur within a process by detecting variation that is inherent in the process and separating it from variation that is changing the process (special causes).

Controllable Poor Quality Costs: These are the costs that management has direct control over to ensure that only acceptable products and services are delivered to the customer. It is divided into two subcategories: prevention costs and nonvalue-added costs.

Cost–Time Charts: A date-and-cost line chart that tracks a process changing cost over time. (Also referred to as a Date-and-Price Chart, and similar to a Gantt Chart.)

Customer-Incurred Poor Quality Costs: These are the costs that the customer incurs when a product or service fails to perform to the customer's expectations. For example, loss of productivity while equipment is down, or travel costs and time spent to return defective merchandise, and the repair cost after the warranty period.

Customer Requirements: These are stated or implied terms that the customer requires to be provided with in order for him or her not to be dissatisfied.

Customer Surveys: Obtaining customers' opinions related to the service or products supplied. This can be done in many ways including phone calls, written surveys, focus groups, one-on-one meetings, etc.

Data Gathering by Document Review: A technique to quickly collect information that currently exists within an organization.

Data Gathering by Interview: The act of using the interviewing process to collect data.

Data Gathering by Samples: A sample is a representation from a population that allows the observer to predict the actual distribution of the population. There is rarely enough time or resources to measure the entire population; a representative sample of the population will yield a model for the entire population.

Data Gathering by Surveys: Surveys are used to measure characteristics of a population relative to such things as behavior, awareness of programs, attitudes or opinions, and needs. Repeated surveys can give valuable information about trends, such as in evaluating government activities.

Data Stratification: A technique used to help identify the underlying causes of variation within a population of data.

Decision-Making Matrix: The team defines the desired results. Then it makes a list of the criteria that are "givens" (must haves) and "wants" (would like to have). The alternative solutions are compared to the "givens" and "wants" list, and a risk analysis is made.

Defects per Million Opportunities (DPMO): This is the average number of errors that would occur in a million opportunities to make an error. It is not the number of defects in a million items. For example, if an item had 10 opportunities for being defective and there were 15 errors in the 1 million units, that would be 1.5 errors per million opportunities. We use errors in place of defects because Six Sigma is now being applied to the service areas where defects are not normally seen, but there are many opportunities to make errors.

Delphi Narrowing Technique: A tool that eliminates the need for face-to-face interaction while it enables achieving group consensus through the use of a prioritization scheme.

Dependent Variable: A variable that we measure as a result of changes in the independent variables.

Design for Manufacturing and Assembly (DFMA): This is a methodology that is used to determine how to design a product for ease of manufacturing. It is usually done by performing concurrent engineering, where manufacturing engineering develops the manufacturing process along with the design.

Design for Six Sigma (DFSS): See DMADV.

Design of Experiment: Structured evaluations designed to yield a maximum amount of information at a defined confidence level at the least expense. They are a set of principles and formulas for creating experiments to define regions of variable value that supports customer satisfaction or to define relations between variables for having more accurate models of phenomena.

Direct Poor Quality Costs: These are costs that can be identified in the organization's ledger.

DMADV: The Six Sigma approach to developing new or redesigning products and/or services. The letter stands for: Define-Measure-Analyze-Design-Verify.

DMAIC: The Six Sigma problem-solving approach. The letters stand for: Define-Measure-Analyze-Improve-Control. It is Six Sigma's version of Shewhart's "Plan, Do, Check, Act" problem analysis technique. Each step in this cycle is designed to ensure best possible results.

Effectiveness: The extent to which an output of a process or subprocess meets the needs and expectations of its customers. A synonym for effectiveness is quality. Effectiveness is having the right output at the right place at the right time at the right price. Effectiveness impacts the customer.

Efficiency: The extent to which resources are minimized and waste is eliminated in the pursuit of effectiveness. Productivity is a measure of efficiency.

Employee Involvement: A technique for unleashing human potential in organizations and involving people in the change process.

Equipment Poor Quality Costs: This is the cost invested in equipment used to measure, accept, or control the products or services, plus the cost of the space the equipment occupies and its maintenance costs. This category also includes any costs related to preparing software to control and operate the equipment.

Error Proofing: Designing the product and the processes so that it is very difficult for errors to occur.

Experiment: A sequence of trials consisting of independent variables set at predesigned levels that lead to measurements and observations of the dependent variables.

External and Internal Customers: All organizations have internal and external customers. The output from any activity within an organization that goes to another individual within the organization has created an internal customer–supplier relationship. The person who receives the input is the internal customer. External customers are individuals or organizations that are not part of the organization that is producing the product. They typically buy the product for themselves or for distribution.

Facilitated Sessions: A meeting in which the leader (facilitator) guides the discussions through a series of steps designed to arrive at a consensus that is acceptable to all participants. A facilitated session helps the participants to define and support mutual goals and objectives.

Failure Mode and Effect Analysis: Identifies potential failures or causes of failures that may occur as a result of process design weaknesses.

Fast Action Solution Technique (FAST): A breakthrough approach that focuses a group's attention on a single process for a one- or two-day meeting to define how the group can improve the process over the next 90 days. Before the end of the meeting, management approves or rejects the proposed improvements.

Five "S"s or Five Pillars: A system designed to bring organization to the workplace. A translation of the original 5S terms from Japanese to English went like this:

- Seiri: Organization
- Seiton: Orderliness
- Seiso: Cleanliness
- Seiketsu: Standardized Cleanup
- Shitsuke: Discipline

In order to assist users of this tool to remember the elements, the original terminology has been retranslated to the following 5Ss.

- Sort
- Set in Order
- Shine
- Standardize
- Sustain

Five Whys (5 Ws): This is a technique to get to the root cause of the problem. It is the practice of asking five times or more why the failure has occurred in order to get to the root cause. Each time an answer is given, you ask why that particular condition occurred.

Flowchart: A method of graphically describing an existing or proposed process by using simple symbols, lines, and words to pictorially display the sequence of activities. Flowcharts are used to understand, analyze, and communicate the activities that make up major processes throughout an organization. They are essential tools used in Process Redesign, Process Reengineering, Six Sigma, and ISO documentation.

Focus Groups: A group of people who have a common experience or interest is brought together where a discussion related to the item being analyzed takes place to define the group's opinion/suggestions related to the item being discussed.

Force Field Analysis: A method to help identify the positive and negative forces working on a process when trying to attain a new state. It is a visual aid for pinpointing and analyzing elements that resist change (restraining forces) or push for change (driving forces).

This technique helps drive improvement by developing plans to overcome the restrainers and make maximum use of the driving forces.

Future State Mapping: This usually takes the form of a flow diagram or a simulation model where a proposed change is drawn out pictorially to better understand the process. In the case where a simulation model is developed, the process can be operated over a period of time based on the assumptions made in the simulation model to determine how effective it will operate.

Gantt Chart: A Gantt chart is a bar chart laid on its side. It is typically used for conveying a project schedule. It is an effective way of identifying interrelationships between tasks and helping to define critical paths through a process or project.

Gap Analysis: A gap analysis is used to compare a present item to a proposed item. It typically will compare efficiency and effectiveness measurements between one product to a competitor's product or one process to another process. It reveals the amount of improvement necessary to bring it in line with the process or product with which it is being compared.

Graphs: A method for visually comparing two or more sets of data. Graphs are visual displays of quantitative or qualitative data. They visually summarize a set of numbers or statistics.

Green Belt: These are individuals who have been trained in the improvement methodologies of Six Sigma and who will be able to lead a Six Sigma process improvement team or work on a process improvement team that is led by a Black Belt or Master Black Belt. This is a part-time job and they maintain their full-time job while performing this activity. They work under the guidance of a Black Belt.

Hard Consensus: When all members of the team absolutely agree with the outcome or solution.

Histograms: A visual representation of the spread or distribution. It is represented by a series of rectangles or "bars" of equal class sizes or width. The height of the "bars" indicates the relative number of data points in each class.

Hoshin Kanri: This in an annual planning process that is used to develop the Hoshin plan or policy development. It is used to set the direction of the improvement activities within the organization. Hoshin is made up of two Chinese words: *Ho*, which means method or form, and *Shin*, which means shiny needle or compass. Kanri means control or management. It is a very systematic, step-by-step planning process that breaks down strategic objectives against daily management tasks and activities.

House of Quality: A matrix format used to organize various data elements, so named for its shape, is the principal tool of quality function deployment (QFD).

Hypothesis Testing: Hypothesis testing refers to the process of using statistical analysis to determine if the observations that differ between two or more samples are caused by random chance or by true differences in the sample. A null hypothesis (Ho) is a stated assumption that there is no difference in the parameters of two or more populations. The alternate hypothesis (Ha) is a statement that the observed differences or relationships between the populations are real and are not the results of chance or an error in the sampling approach.

Independent Variable: An independent variable is an input or process variable that can be set directly to achieve a desired result.

Indirect Cost: These are the costs that are imposed on an output that is not directly related to the cost of the incoming materials or the activities that transform it into an output. It is all the support costs that are needed to run the business that are applied against the product in order to make a profit. For example, the cost of accounting, personnel, ground maintenance, etc.

Indirect Poor Quality Costs: Costs that are incurred by the customer or costs that result from the negative impact poor quality has on future business, or lost opportunity costs.

Initiating Sponsor: Individual/group who has the power to initiate and legitimize the change for all of the affected individuals.

Innovation: Converting ideas into tangible products, services, or processes.

Intangible Benefits: These benefits are gains attributed to an improvement project that are not documented in the formal accounting process. They are often called "soft benefits." Frequently, they are savings that result from preventive action that stops errors from occurring.

Internal Error Costs: The costs incurred by the organization as a result of errors detected before the organization's customer accepts the output. In other words, it is the costs the organization incurs before a product or service is accepted by the customer because someone did not do the job right the first time.

ISO 9000 Series: A group of standards released by the International Organization for Standardization, Zurich, Switzerland, that defines the fundamental building blocks for a Quality Management System and the associated accreditation and registration of the Quality Management System (QMS).

Just-in-Time: A major strategy that allows an organization to produce only what is needed, when it's needed, to satisfy immediate customer requirements. Implemented effectively, the just-in-time concept will almost eliminate in-process stock.

Kaikaku: This is a revolutionary type of activity. While Kaizen is evolutionary, Kaikaku is similar to process reengineering or redesign.

Kaizen: This is a Japanese term that means continuous improvement. *Kai* means change and *zen* means good or for the better.

Kanban: This is usually a printed card that contains specific information related to parts, such as the part name, number, quantity needed, etc. It is the primary communication used in just-in-time manufacturing. It is used to maintain effective flow of materials through an entire manufacturing system while minimizing inventory and work in process. It is used in place of complex production control computer systems.

Kano Model: A theory of product development and customer satisfaction, which classifies customer preferences into five categories:

- Must-Be Quality
- One-Dimensional Quality
- Attractive Quality
- Indifferent Quality
- Reverse Quality

Key Performance Indicators (KPI): KPI stands for "key performance indicators." These measurements indicate the key performance parameters related to a process, organization, or output. They are the key ways by which that item is measured and are usually used to set performance standards and continuous improvement objectives. They are sometimes called CPI (critical performance indicators).

Knowledge Management: This is a system for capturing the knowledge that is contained within an organization. It groups knowledge into two categories. The first classification is tacit knowledge (soft knowledge).

This knowledge is undocumented, intangible factors embodied in an individual's experience. The second classification is explicit knowledge (hard knowledge). This knowledge is documented and quantified.

Lean Manufacturing: This is a focus on eliminating all waste in the manufacturing process. It includes Lean principles, such as:

- Zero inventory
- Batch to flow, cutting batch size
- Line balancing
- Zero wait time
- Pull instead of push production control systems
- Cutting actual process time

Lean Thinking: This is a focus on eliminating all waste within the processes, including customer relations, product design, supplier networks, production management, sales, and marketing. Its objective is to reduce human effort, inventory, cycle time, and space required to produce customer-deliverable outputs.

Line Graph: The simplest graph to prepare and use is the line graph. It shows the relationship of one measurement to another over a period of time. Often this graph is continually created as measurement occurs. This procedure may allow the line graph to serve as a basis for projecting future relationships of the variables being measured.

Machine Capability Index (Cmk): This is a short-term machine capability index derived from observations from uninterrupted production runs. The preferred Cmk value is greater than 1.67. The long-term machine capability index should be greater than 1.33.

Market Segmentation: This occurs when the total market for an individual product or service is subdivided into smaller groups based upon the individual characteristics of the group. This allows different market strategies to be applied to the segmented market areas.

Master Black Belt: Master Black Belts are experts in the Six Sigma methodology and are responsible for the strategic implementation of Six Sigma throughout the organizations. They are responsible for training and mentoring Black Belts and Green Belts. They also are responsible for conducting complex Six Sigma improvement projects and should develop, maintain, and revise the Six Sigma materials. They are responsible for applying statistical controls to difficult problems that are beyond the Black Belts' knowledge base.

Matrix Diagrams: A way to display data to make it easy to visualize and compare.

Mean: This is the average data point value within a data set. It is calculated by adding all of the individual data points together, then dividing that figure by the total number of data points.

Measurement Error: This is the error that is inherent in every measurement that is taken. No measurement is precise. Measurement error can be caused by many factors, including human error, equipment precision, and equipment calibration.

Mind Maps: An unstructured cause-and-effect diagram. Also called mind-flow or Brain Webs.

Mistake Proofing: Mistake proofing is also known as poka-yoke. A poka-yoke is any mechanism in a Lean manufacturing process that helps an equipment operator avoid (*yokeru*) mistakes (*poka*). Its purpose is to eliminate product defects by preventing, correcting, or drawing attention to human errors as they occur.

Motivation Management: It is developing an understanding of what provides an individual with a sense of self-worth, pride, and accomplishment when using this understanding to guide the way work is assigned and how individuals are recognized for their efforts. It results in an organization developing its values, beliefs, procedures, and culture in a way that drives the organization's employees to get more enjoyment out of the work they are doing and be more committed to the organization. It makes use of a set of logical neuro programs that suit the perception of a person's or an organization's needs for the purpose of efficiency and optimality to accomplish desired organizational goal.

Negative Analysis: An approach used to look at a process or situation to define what action could be taken to cause a negative impact upon the results. It generates a list of actions that, if implemented, would result in making the present situation worse. It then generates action plans to minimize the impact that these actions would have upon the process or situation.

Nominal Group Technique (NGT): A special purpose technique, useful for situations where individual judgments must be tapped and combined to arrive at decisions. It is a process to develop and narrow alternatives by generating ideas.

Nonvalue-Added Costs: These are the costs of doing activities that the customer would not want to pay for because it adds no direct value to him or her. It can be further divided into business value-added, no value-added, and bureaucracy costs. It also includes appraisal costs.

Normal Distribution: Occurs when frequency distribution is symmetrical about its mean or average.

Normal Probability Plots: This is used to check whether observations follow a normal distribution. $P > 0.05$ = data is normal.

One Piece Flow: This is a concept where a single piece of work moves between workstations instead of a batch process.

Organizational Change Management: A methodology designed to help prepare the organization and the individuals within the organization to accept changes to the organizational structure, processes, and operating procedures. It is designed to break down resistance to change and to build up organization resiliency.

Organizational Excellence: This methodology is made up of five key elements, called the *Five Pillars*, which must be managed simultaneously to continuously excel. The five pillars include:

- Process Management
- Project Management
- Change Management
- Knowledge Management
- Resource Management

Organizational Process Consultation: It is the combination of skills in establishing a helpful relationship, in knowing what kind of processes to look for, and in intervening in such a way that processes are improved.

Organizational Process Improvement (OPI): A combination of two methodologies, which are process reengineering and process redesign. It is often also called Business Process Improvement (BPI). OPI is a systematic approach to bringing about step function improvement in processes within an organization. It focuses on increasing adaptability, efficiency, and effectiveness while reducing cost and cycle time.

Origin: The point where the two axes on an X–Y graph meet. When numbers are used, their value is increased on both axes as they move away from the origin.

Other Point of View (OPV): A method that aids in idea generation and evaluation by careful examination of the views of stakeholders involved. It is generally more effective when used early in the process, for idea generation as opposed to idea evaluation. It also tends to be more effective with small groups (two or three people) than with larger ones.

Pareto Analysis: A specialized type of column graph that is created to simplify comparisons between items.

Performance Goals: Quantifies the results that will be obtained if the business objectives are satisfactorily met.

Performance Improvement Plan (PIP): A three-year plan designed to align the environment within an organization with a series of vision statements that drive different aspects of the organization's behaviors.

Performance Plan: A contract between management and the employees that define the employees' roles in accomplishing the tactics, and the budget limitations that the employees have placed upon them.

Performance Standard: Defines the acceptable error level of each individual in the organization.

PERT Charts: PERT stands for Program Evaluation Review Technique. This is a methodology that was developed by the U.S. government in the 1950s. It is a project management tool used to schedule, organize, and coordinate tasks within the project. It provides an effective way of determining interdependencies between activities and timing. It allows for the critical path through the project to be readily defined.

Plan–Do–Check–Act: A structured approach for the improvement of services, products, and/or processes developed by Walter Shewhart.

Poisson Distribution: It is an approximation of the binomial when p is equal to or less than 0.1 and the sample size is fairly large (p = probability). It is used as distribution of defect counts and can be used as an approximation of the binomial. It is closely related to the exponential distribution. It is used to model rates, like errors per output, inventory turns, or arrivals per hour.

Policy Deployment: An approach to planning in which organization-wide, long-range objectives are set, taking into account the organization's vision, its long-term plan, the needs of the customers, the competitive and economic situation, and previous results.

Poor Quality Cost (PQC): A methodology that defines and collects costs related to resources that are wasted as a result of the organization's inability to do everything correct every time. It includes both direct and indirect costs.

Portfolio Project Management: This is a technique used to manage all of the projects going on within a specific area. In the past when projects were managed independently, resources were not always assigned in the best manner. This technique optimizes the success of the critical projects that have priority within the organization.

Positive Correlation: This occurs when both variables increase or decrease together. Negative correlation is when one variable increases while the other one decreases.

PPM: This stands for parts per million. Typically, in Six Sigma, it is used for defects per million opportunities (DPMO).

Prevention Costs: These are all the costs expended to prevent errors from being made or, to say it another way, all the costs involved in helping the employee do the job right every time.

Preventive Action: This is action taken that will eliminate the possibility of errors occurring rather than reacting from errors that occurred. It is a long-term, risk-weighted action that prevents problems from occurring based on a detailed understanding of the output and/or the processes that are used to create it. It addresses inadequate conditions that may produce errors.

Prioritization Matrices: A narrowing technique that is used to rank large lists of alternatives.

Process: A series of interrelated activities or tasks that take an input and provide an output.

Process Benchmarking: A systematic way to identify superior processes and practices that are adopted or adapted to a process in order to reduce cost, decrease cycle time, cut inventory, and provide greater satisfaction to the internal and external customers.

Process Capability Analysis (Cp): A statistical comparison of a measurement pattern or distribution to specification limits to determine if a process can consistently deliver products within those limits process. It is a measure of the relationship between common system variation and the specification limit.

Process Control: It is a way the process is designed and executed to maximize the cost effectiveness of the process. It includes process initiation, selection of the process steps, selection of alternative steps, integration of the individual activities into the total process, and the termination of the process. Too frequently, process control and process control charts are used interchangeably and they should not be.

Process Improvement Team: A group of employees assigned to improve a process. It is usually made up of employees from different departments.

Process Owner: The individual responsible for the process design and performance. He/she is responsible for the overall performance from the start of the process to the satisfaction of the customer with the delivered output. It is the responsibility of the process owner to ensure that suboptimization does not occur throughout the process as well as setting improvement performance goals for the process.

Process Qualification: A systematic approach to evaluating a process to determine if it is ready to ship its output to an internal or external customer.

Process Redesign: A methodology that takes the current process and removes the waste out of it and streamlines it. It's used when a 20 to 60 percent reduction in cost or cycle time is required.

Process Reengineering: A methodology used to radically change the way a process is designed by developing an aggressive vision of how it should perform and using a group of enablers to prepare a new process design that is not hampered by the present process's paradigms. Used when a 60 to 90 percent reduction in cost or cycle time is required.

Process Simplification: A methodology that takes complex tasks, activities, and processes and bisects them to define less complex ways of accomplishing the defined results.

Project Charter: A document that formally organizes the project thereby authorizing the project leader to enlist organizational resources to accomplish its objectives. It also defines what the project is responsible for accomplishing.

Project Management: The application of knowledge, skills, tools, and techniques to project activities in order to meet or exceed stakeholders' needs and expectations from a project (From the *Project Management Body of Knowledge* (*PMBOK®*) Guide).

Pull System: This is a production control system that replaces parts and components only when the previous part or component has been consumed. It is designed to eliminate in-process storage and is part of a just-in-time system.

Qualitative Data: It is data related to counting the number of items and cannot be broken down into smaller intervals. It is count rather than measurement data. For example, the number of machines shipped in a specific time period.

Quality Function Deployment: A structured process for taking the "voice of the customer," and translating it into measurable customer requirements into measurable counterpart characteristics, and "deploying" those requirements into every level of the product and manufacturing process design and all customer service processes.

Quality Management: All activities of the overall management function that determine the quality policy, objectives, and responsibilities and implement them by means such as quality planning, quality control, quality assurance, and quality improvement within the Quality Management System (QMS) (ISO 8402: 1994).

Quality Plan: A document setting out the specific quality practices, resources, and sequence of activities relevant to a particular product, project, or contract (ISO 8402: 1994).

Reliability Analysis: This is a technique used to estimate the probability that an item will perform its intended purpose for a specific period of time under specific operating conditions.

Reliability Management System: Designing, analyzing, and controlling the design and manufacturing processes so that there is a high probability of an item performing its function under stated conditions for a specific period of time.

Resource Driver: Describes the basis for assigning cost from an activity cost pool to products or other cost objects.

Resultant Poor Quality Costs: These are the costs that result from errors. These costs are called resultant costs because they are directly related to management decisions made in the Controllable Poor Quality Costs category. It is divided into two subcategories: internal error costs and external error costs.

Reverse Engineering: The process of purchasing, testing, and disassembling competitors' products in order to understand the competitors' design and manufacturing approach, then using this data to improve the organization's products.

Rewards and Recognition: This is action taken to reinforce desired behavior patterns or exceptional accomplishments. Categories of rewards and recognition include:

- Financial compensation
- Monetary awards
- Group/team rewards

- Public personal recognition
- Private personal recognition
- Peer rewards
- Customer rewards
- Organizational awards

Risk Analysis: This is an evaluation of the possibility of suffering harm or loss. A measure of uncertainty. An uncertain event or condition that, if it occurred, might have a positive or negative effect on the organization or the project.

Risk Management: Every project carries with it a certain amount of risk. Risk management is the process of identifying and prioritizing those risks, then both implementing strategies to manage them and designing contingency plans to supplement those strategies should the risk occur.

Robustness: The characteristics of a process output and process design that make it insensitive to the variation in inputs.

Role Mapping: A method of graphically defining the relationship of people relative to their professional relationships, political, and organizational structures that are necessary to the success of the various components of a major change.

Roll-Through Yield (RTY): See First-Time Yield.

Root Cause Analysis: The process of identifying the various causes affecting a particular problem, process, or issue and determining the real reasons that caused the condition.

Run Charts: A graphic display of data used to assess the stability of a process over time, or over a sequence of events (such as the number of batches produced). The run chart is the simplest form of a control chart.

Scatter Diagrams: A graphic tool used to study the relationship between two variables. Scatter Diagrams are used to test for possible cause-and-effect relationships. It does not prove that one variable causes the other, but it does show whether a relationship exists and reveals the character of that relationship.

Seven Basic Tools: These are seven quality improvement tools that all employees should be familiar with and be able to use. They were originally generated by Kaoru Ishikawa, a professor of engineering at Tokyo University and the father of Quality Circles. The seven tools include:

1. Cause-and-Effect Diagrams
2. Check Sheets

3. Control Charts
4. Histograms
5. Pareto Charts
6. Scatter Diagrams
7. Stratification

Simplification Approaches: These are a series of techniques that focus on simplifying the way things are done. It could include things such as:
- Combining similar activities
- Reducing amount of handling
- Eliminating unused data
- Clarifying forms
- Using simple English
- Eliminating nonvalue-added activities
- Evaluating present IT activities to determine if they are necessary
- Evaluating present activities to determine if IT approaches would simplify the total operations

SIPOC: This stands for suppliers, inputs, processes, output, and customers. It is used to help ensure that you remember all the factors when mapping a process.

Six Sigma: Six Sigma is a rigid, systematic methodology that utilizes information (managing by fact) and statistical analysis to measure and improve an organization's performance by identifying and preventing errors. It can be thought of in three parts:
1. Metric: 3.4 defects per million opportunities
2. Methodology: DMAIC/DMSS structured problem-solving tools
3. Philosophy: Reduce variation in the organization and drive decisions based on knowledge of the customer

Six-Sigma Program: A program designed to reduce error rates to a maximum of 3.44 errors per million units, developed by Motorola in the late 1980s.

SMED: This stands for Single Minute Exchange of Die. It is one of the Lean tools and it is a key part of just-in-time programs. It is a methodology used to minimize the amount of time of changing a process over to produce another output.

Solutions Evaluation: A technique used to help review and narrow solutions on the basis of a thorough cost/benefit analysis.

Spider Diagrams/Radar Charts: Used to show or compare one or more sets of data to each other. Often used to indicate the status quo (current state) against the vision (future state).

Stakeholder Analysis Plan: A system to identify "key stakeholders" or individuals that have a stake in the overall success/failure of the process.

Standard Deviation: An estimate of the spread (dispersion) of the total population based upon a sample of the population. Sigma (σ) is the Greek letter used to designate the estimated standard deviation.

Statistical Thinking: This is having a complete situational understanding of a wide range of data where several control factors may be interacting at once to influence an outcome.

Storyboarding: The act of physically structuring the output into a logical arrangement. The ideas, observations, or solutions may be grouped visually according to shared characteristics, dependencies upon one another, or similar means. These groupings show relationships between ideas and provide a starting point for action plans and implementation sequences.

Stratification: The process of classifying data into subgroups based on characteristics or categories. It is the breakdown of the whole (total area of concern) into smaller related subgroups.

Supply Chain Management: This is the flow of items from raw materials to accepted products at the customer location. It is a methodology used to reduce cost, lead times, and inventory, while increasing customer satisfaction.

SWOT Analysis: This stands for Strengths, Weaknesses, Opportunities, and Threat analysis. It is used to help match the organization's resources and capabilities to the competitive environment that exists in their market segment. It is often used as part of the strategic planning process.

System: The organizational structure, responsibilities, procedures, and resources needed to conduct a major function within an organization or to support a common business need.

Tactic: How the strategies will be implemented.

Taguchi Methods: These are design of experiment approaches by Dr. Taguchi for use where the output depends on many factors without having to collect data using all possible combinations of values for these variables. It provides a systematic way of selecting combinations of variables so that their individual effects can be evaluated.

Takt Time: Takt time is the rate that a completed item leaves the last step in the process. It should be equivalent to the rate at which customers, internal or external, require the output. It drives the pull system as it eliminates the need for in-process stock. The process should be designed so that each step in the process is operating at the same takt time as the sales process. This is the ideal situation that keeps the process in continuous flow without buildup within the process or between processes.

Team: A team is a small group of people who work together to realize their interdependencies and understand that both personal and team goals are best accomplished with mutual support.

Theory of Constraints (TOC): This is a set of tools that examines the entire system to define continuous improvement opportunities. It consists of a number of tools. For example:

- Transition Tree
- Prerequisite Tree
- Current Reality Tree
- Conflict Resolution Diagram
- Future Reality Tree

Throughput Yield (TPY): This is the yield that comes out of the end of a process after any errors that are detected have been scrapped or reworked and reentered into the process. Effective rework procedures can often increase first-time yield from 10 percent to a throughput yield of 100 percent.

Time-Related Process Capability Index (Cpk): This takes into account the drift that a product will have over time caused by common variation. Caused by things like difference operators, different setups, allowable differences in material, etc. For all customer impact measurements, a Cpk of at least 1.33 is normal accepted standards unless the product is screened to protect the customer.

Toll Gate: These are process checkpoints where deliverables are reviewed and measured, and readiness to move forward is addressed. Typically, if a total project has not completed all of its commitments that are due at a toll gate, the project does not progress to the next level until these commitments are met.

Typically, this is a management review to determine if the project should continue.

Total Quality Management: A methodology designed to focus an organization's efforts on improving quality of internal and external products and services. ISO 8402 defines it as: A management approach of an organization, centered on quality, based on the participation of all its members and aiming at long-term success through customer satisfaction and benefits to the members of the organization and to society.

Tree Diagrams: A systematic approach that helps the user think about each phase or aspect of solving a problem, reaching a target, or achieving a goal.

TRIZ: This is a methodology that was developed in Russia and stands for "theory of innovative problem-solving." It was originated by Henrich Alashuller in 1946. It is effective at identifying low-cost improvement solutions during the define or identify phase. It is helpful in defining root cause of defects.

Types of Data: There are basically two major groupings of data. They include:

Attributes Data: The kind of data that is counted, not measured. It is collected when all you need to know is yes or no, go or no-go, accept or reject.

Variables Data: Variables data are used to provide a much more accurate measurement than attributes data provide. It involves collecting numeric values that quantify a measurement and, therefore, requires a smaller sample to make a decision.

Types of Teams: There are many different types of teams that are identified by different properties related to the team organization and objectives. Typical teams include:

- Department Improvement Teams, focusing on individual area improvement opportunities
- Quality Circles, voluntary teams that form themselves
- Process Improvement Teams, typically working across functions, focusing on optimizing a total process typified by process reengineering and process redesign
- Task Forces, typified by an emergency that occurs within an organization
- Natural Work Teams, made up of individuals who are brought together to perform ongoing activities.

Value-Added Analysis: A procedure for analyzing every activity within a process, classifying its cost as value-added or nonvalue-added, and then taking positive action to eliminate the nonvalue-added cost.

Value Analysis (VA): The act of identifying the required functions for a product, establishing values for the required functions, and suggesting an approach to provide the required functions at the lowest overall cost without performance loss to optimize cost performance.

Value Engineering (VE): A methodology that seeks to improve the "value" of goods or products and services by evaluating ways that costs or function can be improved without having a negative impact upon the other parameter. Value can be increased by either improving the function of the item or reducing the cost to produce the item. Value, as defined, is the (function)/(cost). Therefore, improving the function or reducing the cost will increase the value. Basic functions must be preserved and not reduced as a consequence of pursuing value improvements. In the United States, value engineering is specifically spelled out in Public Law 104–106, which states, "Each executive agency shall establish and maintain cost-effective value engineering procedures and processes."

Value Proposition: A document that defines the benefits that will result from the implementation of a change or the use of an output as viewed by one or more of the organization's stakeholders. A value proposition can apply to an entire organization, parts thereof, or customers, or products, or services, or internal processes.

Value Stream: This is all of the steps/activities (both value-added, business value-added, and nonvalue-added) in a process that the customer is willing to pay for.

Value Stream Mapping: This tool is used to help you understand the flow of materials and information as an item makes its way through the value stream. A value stream map takes into account not only the item, but also the management and information systems that support the basic item. This is helpful in working with cycle-time reduction problems and is primarily used as part of the Lean tool kit.

Variable Control Charts: A plot of variables data of some parameter of a process' performance, usually determined by regular sampling of the product, service, or process as a function (usually) of time or unit numbers or other chronological variable. This is a frequency distribution plotted continuously over time that gives immediate feedback about the behavior of a process. A control chart will have the following elements:

- Center Line (CL)
- Upper Control Limit (UCL)
- Lower Control Limit (LCL)

Variables Data: The kind of data that is always measured in units, such as inches, feet, volts, amps, ohms, centimeters, etc. Measured data give you detailed knowledge of the system and allow for small, frequent samples to be taken.

Variation: This is a measure of the changes in the output from the process over a period of time. It is typically measured as the average spread of the data around the mean and is sometimes called noise.

Vision: A description of the desired future state of an organization, process, team, or activity.

Vital Few: This is the 20 percent of the independent variables that contribute to 80 percent of the total variation.

Voice of the Business (VOB): This describes the stated and unstated needs and requirements of the organization and its stakeholders.

Voice of the Customer (VOC): This describes the stated and unstated needs and requirements of the external customer.

Voice of the Employee (VOE): This is the term used to describe the stated and unstated needs and requirements of the employees within your organization.

Voice of the Process (VOP): This is the term used to describe what the process is telling you about what it is capable of achieving.

Voting: A method of selecting an item from a list of alternatives. It is a method that allows an individual to identify his or her preference from a group of alternatives.

Work Breakdown Structure (WBS): This is a Gantt chart used in project management to monitor and plan the activities related to doing the project as well as defining their interrelationships and their present status.

Workflow Diagram: This diagram visually represents the movement and transfer of resources, documents, data, and tasks through the entire work process for a given product or service. The diagram is based on the layout of the organization or the department with which the process comes in contact.

Workflow Monitoring: An online computer program that is used to track individual transactions as they move through the process to minimize process variation.

Work Standards: When work standards are practiced, everyone in the organization is committed to performing the work in the same best way. Work standards include documentation methods and developing engineering standards to set expectation and measurement matrix. They provide job aids and training to the employees that effectively communicate the best ways to perform an activity and sets the minimum performance standard for the trained employee.

Yellow Belt (YB): A Yellow Belt typically has a basic understanding of Six Sigma, but does not have the experience, training, or capability to lead projects on his or her own. They work on special assignments to assist Green Belts and Black Belts in developing and implementing Six Sigma projects. As Yellow Belts gain experience, they are good candidates for Green Belt training.

Zero Defects: This was a complete system directed at eliminating all defects from a product. It was originated by Phil Crosby on a military contract and spread throughout the world. It sets a higher standard for performance than Six Sigma by 3.4 defects per million opportunities. It focused on perfection, which is impossible to reach.

Index

D

G

H

T